Sustainable Development Goals Series

The **Sustainable Development Goals Series** is Springer Nature's inaugural cross-imprint book series that addresses and supports the United Nations' seventeen Sustainable Development Goals. The series fosters comprehensive research focused on these global targets and endeavours to address some of society's greatest grand challenges. The SDGs are inherently multidisciplinary, and they bring people working across different fields together and working towards a common goal. In this spirit, the Sustainable Development Goals series is the first at Springer Nature to publish books under both the Springer and Palgrave Macmillan imprints, bringing the strengths of our imprints together.

The Sustainable Development Goals Series is organized into eighteen subseries: one subseries based around each of the seventeen respective Sustainable Development Goals, and an eighteenth subseries, "Connecting the Goals", which serves as a home for volumes addressing multiple goals or studying the SDGs as a whole. Each subseries is guided by an expert Subseries Advisor with years or decades of experience studying and addressing core components of their respective Goal.

The SDG Series has a remit as broad as the SDGs themselves, and contributions are welcome from scientists, academics, policymakers, and researchers working in fields related to any of the seventeen goals. If you are interested in contributing a monograph or curated volume to the series, please contact the Publishers: Zachary Romano [Springer; zachary.romano@springer.com] and Rachael Ballard [Palgrave Macmillan; rachael.ballard@palgrave.com].

Nicolette V. Roman · Shewli Kumar ·
Rashila Ramli · Beatrix Schwarzer
Editors

Social Inequalities Within and Among Countries

Achieving SDG 10 Through Social Interventions

Springer

Editors
Nicolette V. Roman
Centre for Interdisciplinary Studies of Children, Families and Society
University of the Western Cape
Belville, South Africa

Rashila Ramli
Institute of Graduate Studies
Universiti Poly-Tech Malaysia
Kuala Lumpur, Malaysia

Shewli Kumar
Department of Social Work
Tata Institute of Social Sciences
Mumbai, Maharashtra, India

Beatrix Schwarzer
Department of Social Work
Frankfurt University of Applied Sciences
Frankfurt, Germany

ISSN 2523-3084 ISSN 2523-3092 (electronic)
Sustainable Development Goals Series
ISBN 978-3-032-08740-9 ISBN 978-3-032-08741-6 (eBook)
https://doi.org/10.1007/978-3-032-08741-6

The German Academic Exchange Service (DAAD) provided funding for the project conducted in Germany, Malaysia, India and South Africa.

This Springer imprint is published by the registered company Springer Nature Switzerland AG
The registered company address is: Gewerbestrasse 11, 6330 Cham, Switzerland

Foreword

I count it a great privilege to write the foreword for this timely publication on social inequalities within and among countries based on SDG 10. It is impressive that 34 academics through 18 chapters have articulated the challenges and achievements in addressing social inequalities in four countries, namely South Africa, Germany, India, and Malaysia.

This book is significant for three reasons:

Firstly, the theory of poverty, inequality, and development is taken from the SDGs which is holistic impacting economic, social, environment, and effective delivery based on addressing people's participation in decision-making and good governance. It is not just income measurement but multi-dimensional. While SDG 1 addresses poverty and moving people up the socio-economic ladder from a minimum per day income measurement. SDG 2 on hunger addresses issues pertaining to malnutrition and related concerns such as stunting. The correlations to poverty and gender is the focus in SDG 5 however on health and education is addressed respectively via SDG 3 and SDG 4. Being in a neighbourhood and housing is the focus of SDG 11.1 which is on affordable housing. By way of theory, public policy research and analytical tools, a multi-dimensional approach is imperative to capture the root causes, experience of deprivation and every day lived realities.

Second, the development agenda is based on human rights and is fundamental as poverty and inequality have many faces especially that linked to discrimination, lack of equal opportunities, and the failure to reduce inequality of outcomes, a theme highlighted in SDG 10.3. Therefore, addressing discrimination through promoting and enforcing non-discriminatory laws and policies is key as emphasised in SDG 16.b. Leaving no one behind is the definite goal. The variables for measurement and monitoring encompass gender, ethnicity, class, socio-economic status, caste, disability as well as locational rural-urban divide including between citizens and stateless or migrant communities or across countries. This aspect is clearly a target as found in SDG 10.2.

Here we can draw from the writing of Amartya Sen who articulated development as freedom and noted that "greater freedom enhances the ability of people to help themselves".[1] It is the "freedom of people to act as they like in deciding on where to work, what to produce, what to consume and so on".[2]

---

[1] Amartya Sen (1999) Development as freedom. (Alfred A. Knopf: New York) pg. 18.
[2] Ibid pg. 27.

Third, analysis and advocacy must lead to effective intervention and problem solving. It is not just improvement in socio-economic dimensions but also in human wellbeing namely an overall improvement in the quality of life. Here the SDG goals, targets, and indicators are not aspirational, it's for effective implementation. While it is the role of the state as duty bearers in the SDGs but based on SDG 17.17 it is a multi-stakeholder intervention model. A shared responsibility between the government, the people, private sector, and non-profit/voluntary sector is emphasised. The modalities for greater role of local communities, the non-profit sector as well as the business community needs further strengthening especially civil society and social enterprises as partners in the delivery system.

Grassroot communities are innovating especially those who have adopted a social solidarity economy (SSE) model in empowering local economies. These communities are undertaking economic activities but at the same time valuing people over profits, ensuring inclusive development, enhancing conservation and breaking down barriers towards greater justice and fairness. In recent times the United Nations[3] through two SSE resolutions and a Secretary General SSE report has confirmed how impactful and empowering these community models are towards the acceleration of SDGs. In this book through the ground experiences, we are able to also articulate not just good theory, analysis but alternative grassroots experiences which are setting new paradigms in addressing socio-economic inequalities.

I take this opportunity to congratulate the editors and writers and call upon all to read, study, and most of all draw out development policy implications which can enhance effective delivery and in so doing, ensure no one is left behind.

Prof. Datuk Dr. Denison Jayasooria
Head of Secretariat All Party
Parliamentary Group Malaysia
on SDGs, Chair of the Society
for the Promotions of SDGs
and Former Professor of Public
advocacy Institute of Ethnic Studies
National University
of Malaysia (UKM)
Bangi, Malaysia

[3]https://appgm-sdg.com/wp-content/uploads/2025/05/Social-Solidarity-Economy-as-a-Vehicle-for-Building-Local-Economies-and-Accelerating-the-Loc.pdf.

Preface

In a world characterised by ongoing inequality and unprecedented global problems, from rising socio-economic inequalities to the lingering effects of a pandemic, the need to tackle systemic inequalities has never been more essential and pressing. The Sustainable Development Goals are humanity's shared desire and aspiration for a more just future—a shared commitment to create a more fair and sustainable future. A clarion call to tear down barriers based on race, gender, class, geography, and other intersecting identities, SDG 10, "Reducing inequalities within and among countries", calls for action. Focussing on SDG10, this book results from an extraordinary transnational cooperation among four countries—India, South Africa, Germany, and Malaysia—each with unique historical paths, socio-economic realities, and views on inequality. Each of these nations has different historical paths, socio-political settings, and development issues that provide rich insight into the complexity of inequality.

Our scholarly endeavour commenced with a fundamental yet significant acknowledgment: inequality transcends national boundaries, constituting a global predicament that necessitates synchronised efforts across international frontiers. Each of the four nations we represent illustrates distinct facets of this predicament. South Africa remains engaged in confronting the historical ramifications of apartheid while striving to cultivate an inclusive society. India grapples with the intricate interplay of caste, class, and regional inequalities prevailing within its extensive populace. Germany navigates the dual imperatives of economic advancement and social cohesion in a society characterized by increasing diversity. Malaysia endeavours to reconcile development across various ethnic groups and geographical landscapes. The main idea that connects our efforts is understanding that inequality shows up in different ways based on identity factors like gender, race, ethnicity, disability, and economic status, creating specific challenges that can't be solved with simple solutions. Through our joint research initiatives and interventions, we have observed the enriching impact of transnational knowledge exchange on our comprehension of inequality, thereby broadening our arsenal for addressing this issue. By integrating case studies from both the Global North and South, this book emphasises the global significance of Sustainable Development Goal 10 while simultaneously acknowledging the localised contexts that influence its execution.

At its essence, this manuscript is fundamentally grounded in the principle of intersectionality, a theoretical construct developed by Kimberlé

Crenshaw that elucidates how interlocking systems of oppression exacerbate marginalisation. Through this analytical framework, we explore how inequalities are manifested across various dimensions such as gender, race, caste, disability, and socio-economic status, and how these intersections necessitate intricate, contextually relevant interventions. The chapters engage in a critical examination of the philosophical underpinnings of the Sustainable Development Goals (SDGs), recognising legitimate critiques regarding their Western-centric foundations, gaps in implementation, and the inherent conflict between economic imperatives and ecological sustainability. Yet, they also acknowledge the Goals' potential as a catalyst for transformative change when coupled with grassroots activism, transnational solidarity, and rigorous policy coherence, while prioritising human dignity and collective wellbeing.

The COVID-19 pandemic, which exacerbated existing disparities, serves as a stark reminder of the fragility of progress. As the data in this book reveal, the pandemic widened gaps in health, education, and income across all four nations, disproportionately affecting women, migrants, and marginalised communities. These findings underscore the need for resilient, inclusive systems that prioritise equity in times of crisis and calm alike.

This book represents more than academic discourse; it documents real interventions that have touched lives across four continents. From addressing discriminatory practices to facilitating safe migration, from building sustainable livelihoods to creating pathways of opportunity, our work demonstrates that progress toward equality is possible when we combine local knowledge with global perspectives. Structured into four sections, the book moves from conceptual foundations to empirical analyses, policy critiques, and forward-looking strategies. It is designed for scholars, policymakers, social workers, and advocates seeking to understand inequality through a global lens while grounding their work in local contexts.

This project would not have been possible without the dedication of a transnational network of scholars, practitioners, and institutions. We extend our gratitude to the contributors from the University of the Western Cape (South Africa), Universiti Kebangsaan Malaysia, Tata Institute of Social Sciences (India), and Frankfurt University of Applied Sciences (Germany), whose interdisciplinary expertise bridges academia and activism. Our work exemplifies the power of collaborative knowledge production in challenging hegemonic paradigms and amplifying marginalised voices.

As we stand at the midpoint of the 2030 Agenda, we offer this book not as a declaration of victory but as evidence of possibility. The path towards reduced inequalities within and among countries remains challenging, but through continued collaboration, evidence-based interventions, and unwavering commitment to justice, we can build societies where no one is left behind. This book is both a testament to what has been achieved and a call to address what remains undone. The road to equity is fraught with challenges, but it is one we must walk together—across disciplines, borders, and divides. We hope this book inspires not only reflection but also action—urging readers to confront inequities with intersectional rigour, transnational empathy, and an unwavering commitment to justice.

Belville, South Africa
Mumbai, India
Kuala Lumpur, Malaysia
Frankfurt, Germany
May 2025

Nicolette V. Roman
Shewli Kumar
Rashila Ramli
Beatrix Schwarzer

Acknowledgements

Professor Dr. Dagmar Oberlies was the driver of this project, Inequality Within and Among Countries (IWAAC-10). As the book's editors and authors, we would like to sincerely thank her for starting and seeing this project through to its conclusion. An ambitious idea has been turned into a real contribution to the global conversation on inequality thanks to her visionary leadership. With unwavering dedication and intellectual rigour, Prof. Oberlies not only conceived of this transnational collaboration but also fostered it through challenges. She was able to create a special environment where knowledge could freely flow across borders by bridging academic traditions across four different countries. Under her direction, this project became a living laboratory of international collaboration, surpassing traditional academic boundaries. We are particularly appreciative of the many opportunities she established, including arranging workshops that inspired fresh perspectives, mentoring upcoming researchers who will continue this work, and facilitating research exchanges that enabled scholars to immerse themselves in various contexts. The next generation of transnational scholars dedicated to tackling inequality has been sown by the student mobility initiatives she supported. Beyond this publication, Prof. Oberlies' commitment to building institutional capacity has left a lasting legacy. She embodied the very equality ideals this work promotes by insisting on equitable partnership models, which challenged established power dynamics in international academic collaboration. We extend our sincere gratitude for her unwavering support, intellectual generosity, and strong conviction in the ability of collaborative scholarship to tackle global issues. Her vision of a more equitable world created via collective action and shared knowledge is reflected in this book.

The partnership between four universities has transformed how we understand inequality and collaboration. The Frankfurt University of Applied Sciences in Germany became the northern anchor, with Prof. Dagmar Oberlies and Dr. Beatrix Schwarzer providing intellectual rigour and hospitality.

Professor Nicolette Roman at the University of the Western Cape in South Africa opened her campus and community networks to visiting researchers, transforming their research approach. Professor Catherina Schenck's expertise in participatory methodologies challenged their assumptions and challenged them to decolonise knowledge practices. Professor Shewli Kumar at the Tata Institute in India infused their work with, field visits to women's cooperatives and community organisations. Professor Rashila Ramli at Universiti Kebangsaan Malaysia created bridges between different knowledge traditions, navigating cultural nuances and introducing them to local cuisines. The partnership has weathered funding uncertainties, political changes, a global pandemic, and everyday challenges of international collaboration. The partnership has emerged with not just a completed project but strengthened relationships, making it our greatest achievement.

This project would not be possible without the funding support of the German Academic Exchange Service (DAAD). We are grateful for the opportunities of growth this funding has provided.

To each and every author in this book, we thank you for taking up the call and working through the different revisions.

To our editor, Lee-Anne Roux, we are so grateful for your assistance with the finalization of the book. THANK YOU!

Contents

Abbreviations

AA	Asylum Act
AGG	Equal Treatment Act
AIECCE	Asian Institute for Early Childhood Care and Education
APPGM-SDG	All-Party Parliamentary Group Malaysia for Sustainable Development Goals
ASEAN	Association of Southeast Asian Nations
AU	African Union
AWHHC	Agri-Worker Household Census
BAMF	Federal Office of Migration and Refugees
BLLRO	Block Land and Land Reforms Offices
BMUB	Federal Ministry for the Environment, Nature Conservation and Nuclear Safety
BMZ	Federal Ministry for Economic Cooperation and Development
BVLF	Bernard van Leer Foundation
CCT	Conditional Cash Transfer
CEDAW	Convention on the Elimination of All Forms of Discrimination Against Women
CHE	Council on Higher Education
CHETNA	Centre for Health, Education, Training and Nutrition
CJA	Child Justice Act
CoGTA	Co-operative Governance and Traditional Affairs
CRC	Convention on the Rights of the Child
CRPD	Convention on the Rights of Persons with Disabilities
CSG	Child Support Grant
CSR	Corporate Social Responsibility
CYCC	Child and Youth Care Centre
DBSH	German Association of Social Workers
DPA	Deutsche Presse-Agentur
ECD	Early Childhood Development
EFA	Education for All
EPU	Economic Planning Unit
EU	European Union
EWRF	Educational, Welfare, and Research Foundation
FAS	Foetal Alcohol Syndrome
FASD	Foetal Alcohol Spectrum Disorders

FRA	Forest Right Act/Agentur der Europäischen Union für Grundrechte
GBV	Gender-based Violence
GII	Gender Inequality Index
GSSWA	Global Social Service Workforce Alliance
HDI	Human Development Index
HEI	Higher Education Institution
HESI	Higher Education Sustainability Initiative
HLFP	High-Level Political Forum
HSA	Hindu Succession Act
IAB	Institute for Employment Research
IAEG-SDGs	Inter-agency and Expert Group on SDG Indicators
IASSW	International Association of Schools of Social Work
IDC	Immigration Detention Centre
IFSW	International Federation of Social Workers
IHDI	Inequality-adjusted Human Development Index
IKMAS	The Institute of Malaysian and International Studies
ILO	International Labour Organisation
IMF	International Monetary Fund
IOM	International Organisation for Migration
IRDAI	Insurance Regulatory and Development Authority
IT	Information and Technology
IWAAC	Inequality Within and Among Countries
LPG	Liberalisation, Privatisation, and Globalisation
M&E	Monitoring and Evaluation
MASW	Malaysian Association of Social Workers
MCO	Movement Control Orders
MDG	Millennium Development Goal
MDGR	Millennium Development Goals Report
MGNREGS	Mahatma Gandhi National Rural Employment Guarantee Scheme
MIPEX	Migrant Integration Policy Index
MoSPI	Ministry of Statistics and Programme Implementation
MPI	Multidimensional Poverty Index
MQA	Malaysian Qualification Agency
NASW: SA	National Association of Social Workers: South Africa
NCSW	National Council of Social Work
NDP	National Development Plan
NEET	Not in Employment, Education, or Training
NGO	Non-Governmental Organisation
NLFCS	National Land Finance Cooperative Society
NPO	Non-Profit Organisation
NQF	National Qualifications Framework
NTFP	Non-Timber Forest Produce
OECD	Organisation for Economic Co-operation and Development
OHCHR	Office of the High Commissioner for Human Rights
OOPE	Out-of-Pocket Expenses

OWG	Open Working Group
PAR	Participatory Action Research
PBO	Public Benefit Organisation
PCOD	Polycystic Ovarian Disease
PCOS	Polycystic Ovary Syndrome
PPP	Public-Private Partnership
PRA	Participatory Rural Appraisal
PVTG	Particularly Vulnerable Tribal Group
RA	Residence Act
RDI	Rural Development Institute
RI	Revenue Inspector
RSA	Republic of South Africa
SACSSP	South African Council for Social Service Professions
SA-PPA	South African Participatory Poverty Assessment
SAQA	South African Qualifications Authority
SASSA	South African Social Security Agency
SDG	Sustainable Development Goal
SGB	Social Codes
SHG	Self-Help Group
SSA	Sub-Saharan Africa
STD	Sexually Transmitted Disease
TAP	Transparency, Accountability and Participation
THE	Times Higher Education
TISS	Tata Institute of Social Sciences
ToC	Theory of Change
UAS	University of Applied Sciences
UK	United Kingdom
UKM	Universiti Kebangsaan Malaysia (National University of Malaysia)
UN	United Nations
UNCED	United Nations Conference on Environment and Development
UNDESA	United Nations Department of Economic and Social Affairs
UNDP	United Nations Development Programme
UNECA	United Nations Economic Commission for Africa
UNECLAC	United Nations Economic Commission for Latin America and the Caribbean
UNESCAP	United Nations Economic and Social Commission for Asia and the Pacific
UNESCO	United Nations Educational, Scientific and Cultural Organisation
UNFPA	United Nations Population Fund
UNGA	United Nations General Assembly
UNHCR	United Nations High Commissioner for Refugees
UNHRC	United Nations Human Rights Council
UNICEF	United Nations Children's Fund

UNRIC	United Nations Regional Information Centre for Western Europe
UNSD	United Nations Statistics Division
UPTM	Universiti Poly-Tech Malaysia
US	United States
USA	United States of America
USAID	United States Agency for International Development
UWC	University of the Western Cape
VNR	Voluntary National Review
WBLRA	West Bengal Land Reforms Act
WEF	World Economic Forum
WFP	Women on Farms Project
WHO	World Health Organisation
WTO	World Trade Organisation

An Introduction to the Sustainable Development Goals (SDGs)

1

Nicolette Roman, Rashila Ramli, Muhammad Rahimi Hasan, Shewli Kumar, and Beatrix Schwarzer

Abstract

The Sustainable Development Goals (SDGs), particularly SDG 10, are intended to alleviate inequality within and between countries. This chapter presents an overview of the key concepts of social justice, intersectionality, and social inequality, as well as the connections between these frameworks in terms of global development goals. This chapter emphasises the importance of multidisciplinary investigation in understanding the complexities of inequality by examining the social, economic, and political aspects that contribute to disparities. It examines what happened in four countries: South Africa, Germany, India, and Malaysia. This demonstrates that different strategies and promises are required to make the SDGs effective in each country. This chapter also discusses the importance of international academic networks in fostering collaboration, knowledge sharing, and policy influence in order to make the world a more equitable place. The opportunities and problems mentioned in this introduction are recommended s to be explored later in the book, when it discusses how social interventions might assist in achieving social justice and equity.

Keywords

Multidisciplinary · Inequality · Intersectionality · Transnational

1.1 Introduction

Inequalities emerge from the socio-economic and political conditions in which individuals live and interact. These disparities are rooted in differential access to resources necessary for individuals, communities, and societies to thrive and progress towards a life of dignity and general well-being. Inequalities undermine individuals, social groups, and communities, violate their rights to reasonable resources, and often lead to violence.

Intersectionality (Crenshaw 1989) offers a theoretical framework for understanding the multiple, overlapping dimensions of inequality.

N. Roman (✉)
University of the Western Cape, Bellville, South Africa
e-mail: nroman@uwc.ac.za

R. Ramli
Universiti Poly-Tech Malaysia, Kuala Lumpur, Malaysia
e-mail: rashilaramli@uptm.edu.my

M. R. Hasan
Universiti Kebangsaan Malaysia, Bangi, Selangor, Malaysia
e-mail: rahimihasan@ukm.edu.my

S. Kumar
School of Social Work, Tata Institute of Social Sciences, Deonar, Mumbai, Maharashtra, India
e-mail: shewli4@tiss.ac.in

B. Schwarzer
University of Frankfurt of Applied Sciences, Frankfurt am Main, Germany
e-mail: schwarzer@fb4.fra-uas.de

N. V. Roman et al. (eds.), *Social Inequalities Within and Among Countries*, Sustainable Development Goals Series, https://doi.org/10.1007/978-3-032-08741-6_1

Applying an intersectional analysis enables a deeper understanding of the barriers individuals and groups face—whether based on gender, race, class, ethnicity, geography (rural-urban), region, disability, or other factors. It identifies the specific interweaving of inequalities experienced in a given context, highlighting the complexities of positionality within social power dynamics and making visible the interconnectedness of social structures. Such analysis provides detailed insights into social constructions and supports the development of targeted social interventions. These interventions aim to promote the dignity and worth of marginalised groups, foster their active participation, and inform their actions to create a more equitable and just world.

National development strategies profoundly impact citizens' livelihoods and well-being. While countries pursue diverse development approaches, they are all influenced by global challenges and frameworks. International agreements, treaties, and development plans address critical issues such as poverty, inequality, and human well-being. Instruments like the Human Development Index (HDI), introduced in 1990, have enabled scholars to track progress over time, while global guidelines, such as the Millennium Development Goals (MDGs) established in 2000, have shaped development priorities.

In 2015, the Sustainable Development Goals (SDGs) replaced the MDGs, setting a new global agenda for development through the 2030 Agenda for Sustainable Development. This agenda, created through a bottom-up consultation process, presents a universal, integrated, and human rights-based framework for sustainable development. It balances economic growth, social justice, and environmental stewardship, emphasising the interconnectedness of peace, development, and human rights.

This chapter introduces a book focused on SDG 10, which has evolved from a project conducted across four countries: Malaysia, India, Germany, and South Africa. The project explored various strategies and challenges involved in addressing inequalities through social interventions both within and across these nations. The examples presented illustrate the diverse strategies and levels of commitment required to localise and achieve the goals.The chapter further highlights the importance of intersectionality, intersectional analysis, and transnational research in addressing inequalities. It underscores the critical role of transnational academic networks in knowledge exchange and policy influence, championing a more inclusive and just global society. By anchoring each subsequent chapter in the ten target areas of SDG 10, the book seeks to provide a comprehensive understanding of the complexities of inequality and the strategies needed to foster a more equitable future.

1.2 History of the Formulation of the SDGs

The formulation of the SDGs is rooted in a complex history spanning decades of international dialogue, cooperation, and action on a range of global challenges. Understanding this history requires tracing back to the early discussions on sustainable development and the evolution of the global development goals. Several key conventions served as major milestones in this process:

1. **Rio Earth Summit (1992)**: The United Nations Conference on Environment and Development (UNCED), also known as the Rio Earth Summit, marked a significant moment in the global environmental movement. It brought together world leaders, policymakers, and civil society to address pressing environmental issues, including climate change, biodiversity loss, and sustainable development. The summit produced several key agreements, including the Rio Declaration on Environment and Development and the Agenda 21 action plan.
2. **MDGs**: In 2000, world leaders gathered at the United Nations Millennium Summit and adopted the Millennium Declaration, which outlined eight MDGs aimed at addressing various dimensions of poverty, health, education, gender equality, and environmental sustainability by 2015. Although the MDGs

focused primarily on social and economic development, they laid critical groundwork for future discussions on sustainable development.

3. **Rio + 20 Conference (2012)**: Twenty years after the Rio Earth Summit, the United Nations Conference on Sustainable Development, also known as Rio + 20, took place in Rio de Janeiro, Brazil. The conference aimed to secure renewed political commitment to sustainable development and to assess the progress made since the original summit. A key outcome of Rio + 20 was the agreement to develop a set of SDGs to build upon the MDGs and to address emerging global challenges.
4. **Open Working Group on Sustainable Development Goals (OWG)**: Following Rio + 20, the United Nations (UN) established the OWG on the SDGs in 2013 to develop a proposal for the post-2015 development agenda. The OWG consisted of representatives from 70 countries who deliberated on various issues related to sustainable development and proposed a comprehensive list of goals and targets.
5. **United Nations General Assembly (UNGA) negotiations**: The OWG's proposal served as the basis for negotiations within the UNGA. Member states engaged in extensive negotiations to refine the goals and targets, considering a range of perspectives and priorities. This inclusive process involved consultations with civil society, the private sector, and other stakeholders.
6. **Adoption of the SDGs (2015)**: In September 2015, world leaders gathered at the UN Sustainable Development Summit in New York and adopted the 2030 Agenda for Sustainable Development, which includes 17 SDGs and 169 associated targets. The SDGs address a wide range of interconnected issues, including poverty, inequality, climate change, environmental degradation, peace, and justice.

Overall, the formulation of the SDGs represents the culmination of decades of international efforts to promote sustainable development and to address global challenges in an integrated and comprehensive manner. The SDGs provide a shared blueprint for a more sustainable and equitable future, requiring collaborative action at local, national, and global levels.

1.2.1 Controversies Surrounding the Philosophy of the SDGs

The SDGs have been subject to constructive criticism, particularly concerning their philosophical foundations. While the SDGs seek to offer a universal framework for global development, critics argue that their implementation must navigate the tension between universalist aspirations and the diverse local contexts of member states (Arora-Jonsson 2023; Pecquet 2023). This contradiction is further highlighted by the fact that although high-level meetings often prioritise macro-level concerns, the SDGs are intended to serve grassroots interests, creating a gap between global agendas and local realities.

Arora-Jonsson (2023) critiques the SDGs for being rooted in Western, neoliberal ideology, advocating for the decolonisation of the development agenda to better reflect diverse cultural and contextual perspectives. Similarly, Pecquet (2023) raises several concerns, including the non-binding nature of the SDGs, inadequate funding, a lack of urgency in their implementation, and the vagueness of their guidelines and indicators. Pecquet further highlights issues such as financial implications, potential political manipulation, insufficient community empowerment, and the paradox of pursuing global equality while struggling to secure commitments from all stakeholders.

Hickel (2019) additionally points to an inherent contradiction between the pursuit of economic growth (SDG 8) and the ecological goals (SDGs 6, 12, 13, 14, and 15), arguing that prioritising growth may undermine environmental sustainability. Despite these controversies, the SDGs were endorsed by 193 member states in 2015 as part of the 2030 Agenda for Sustainable Development, reflecting a shared commitment to their objectives even as debates regarding their implementation and effectiveness persist.

Examining how individual countries have approached the SDGs offers valuable insights into both the challenges and achievements of the agenda. Focusing on the experiences of Germany, India, Malaysia, and South Africa provides a nuanced understanding of the SDGs in practice.

1.2.2 Implementing the SDGs: Experiences from Germany, India, Malaysia, and South Africa

A growing body of literature on the implementation of the SDGs across the four countries includes studies on Germany by Scholz et al. (2016) and Jossin and Peters (2021); India by Bhanja and Roychowdury (2020) and Tiwari et al. (2021); Malaysia by Daud and Ramli (2022, 2024); and South Africa by Mahlatsi (2021) and Ngubane and Pillay (2023).

As explained by Scholz et al. (2016) and Jossin and Peters (2021), the case of Germany shows that implementing the SDGs within the sustainable development framework requires policy coherence under Germany's Sustainable Development Strategy.

The Federal Ministry for the Environment, Nature Conservation and Nuclear Safety (BMUB) and the Federal Ministry for Economic Cooperation and Development (BMZ) are the primary agencies responsible for implementing SDG practices at the ministerial level in Germany, while the Federal Statistical Office is tasked with producing the biennial report on SDG progress. Active participation, monitoring, and review are promoted through public forums, dialogue initiatives, and the engagement of civil society actors (such as trade unions and private sector associations), as well as, to some extent, the Club of 2030 Agenda Municipalities.

In Malaysia, the implementation of the SDG Agenda 2030 involves commitment from the government, civil society, and academia. Daud and Ramli (2022, 2024) note that localisation of the SDGs is largely driven by the All-Party Parliamentary Group Malaysia for Sustainable Development Goals (APPGM-SDG). Localisation operates at three levels: the implementation, monitoring, and evaluation of community projects identified within parliamentary constituencies by yearly cohorts from 2020 to 2030. The number of projects continues to grow, with 34 projects in the 2020 cohort, 87 in 2021, and 193 in 2022, funded under the auspices of APPGM-SDG as well as by the Economic Planning Unit (EPU) in the Prime Minister's Department and the Ministry of Finance.

In India, responsibility for SDG implementation rests with the national government. The National Institution for Transforming India (NITI Aayog) has been designated to oversee monitoring, supported by the Ministry of Statistics and Programme Implementation (MoSPI). Monitoring occurs across various levels—states or union territories, districts, and localities. In 2018, a ranking system was introduced to assess the performance of all states and union territories across all SDGs. To date, SDG implementation and monitoring have been carried out intensively at the state level, with comparatively limited efforts at the city level (Tiwari et al. 2021).

South Africa has made significant efforts to localise the SDGs at the municipal level. According to Ngubane and Pillay (2023), localisation at the municipal level can be understood across four categories: first, some municipalities have developed institutional frameworks specifically for the SDGs; second, most municipalities have initiated activities supporting SDG localisation; third, certain municipalities have aligned their processes and plans with the SDGs; and fourth, all municipalities have at least initiated considerations for integrating Fourth Industrial Revolution (4IR) elements into their localisation strategies (Ngubane and Pillay 2023:10). Figure 1.1 indicates the 17 SDG goals. There is a total of 169 targets and 244 indicators.

The adoption of the SDGs in 2015 marked a pivotal moment in global development efforts, aiming to address multifaceted challenges and promote sustainable development by 2030. According to the United Nations Development Programme (UNDP), significant strides have been achieved in poverty reduction, access to education and healthcare, gender equality, and environmental sustainability. As shown in Table 1.1, the core principles of the SDGs are universality, interconnectedness and indivisibility, inclusiveness,

Fig. 1.1 Goals: 17, Targets: 169, Indicators: 244

Table 1.1 Core principles

1	Universality	Agenda 2030 is applicable in all countries, in all contexts, and at all times
2	Interconnectedness and indivisibility	It is crucial that all entities responsible for the implementation of the SDGs treat them as an integrated whole, rather than approaching them as a menu of individual goals to pick and choose from
3	Inclusiveness	Calls for the participation of all segments of society—regardless of race, gender, ethnicity, and identity—to contribute to its implementation
4	Multi-stakeholder partnerships	Calls for establishing multi-stakeholder partnerships to mobilise and share knowledge, expertise, technology and financial resources

and multi-stakeholder partnerships. These principles uphold the central commitment to "leave no one behind". The SDGs seek to benefit all people, committing to reach those in need and deprivation wherever they are, by specifically targeting their distinct challenges and vulnerabilities.

In practice, applying these principles demands a constant focus on identifying and supporting groups and individuals who are typically excluded or marginalised within society.

It is important to bear in mind that the 17 SDGs should not be studied in isolation. The SDGs have been thematised into five categories. As depicted in the diagram below, the five categories are People, Planet, Prosperity, Peace, and Partnership. In the context of this book, we argue that Gender Equality, while remaining a stand-alone goal, is as dominant as Peace and Partnership.

As shown in Fig. 1.2, the five Ps provide a framework for understanding the SDGs. By grouping the SDGs into five main categories, this framework highlights the interlinkages between them. As illustrated in the diagram below, Peace and Partnership are two categories that are present in all other SDGs. Another SDG that can be found across all the SDGs is SDG 5 on Gender Equality. This is because the participation of women, men, girls, and boys in all activities will ensure a more inclusive society. Issues of access to resources and opportunities must be addressed across all SDGs.

1.3 Understanding SDG 10: Inequalities Within and Among Countries

SDG 10, one of the 17 SDGs outlined in the 2030 Agenda for Sustainable Development, was adopted by all United Nations Member States in 2015. It focuses on "reduced inequalities"and aims to address various forms of inequality within and among countries. This goal is essential for promoting inclusive and sustainable development. Addressing inequalities requires concerted

Fig. 1.2 Five categories of the SDGs: people, planet, prosperity, peace, and partnership

efforts from governments, civil society, the private sector, and the international community. By implementing targeted policies and initiatives, fostering inclusive societies, and promoting equitable access to opportunities, progress can be made towards achieving SDG 10 and creating a more just and prosperous world for all.

SDG 10 consists of 7 targets with three expanded ones. That aim to reduce income inequality, enhance social, economic, and political inclusion, ensure equal opportunities, and empower marginalised and vulnerable groups.

1. **Income Inequality**: Reduce income inequality within and among countries.
2. **Social Inclusion**: Empower and promote the social, economic, and political inclusion of all, irrespective of age, sex, disability, race, ethnicity, origin, religion, economic or other status.
3. **Equality of Opportunity**: Ensure equal opportunity and reduce inequalities of outcome, including by eliminating discriminatory laws, policies, and practices and promoting appropriate legislation, policies, and action in this regard.
4. **Progressive Policies**: Adopt policies, especially fiscal, wage, and social protection policies, and progressively achieve greater equality.
5. **Regulation of Global Financial Markets**: Improve the regulation and monitoring of global financial markets and institutions and strengthen the implementation of such regulations.
6. **Empowerment of Inclusive Societies**: Ensure enhanced representation and voice for developing countries in decision-making in global international economic and financial institutions.
7. **Facilitating Migration**: Facilitate orderly, safe and responsible migration and mobility of people. There are three expanded dimensions in this category.

SDG 10 is crucial for fostering inclusive and sustainable development. Addressing inequalities within and among countries is essential for achieving social cohesion, economic growth, and environmental sustainability. Persistent inequalities in income, wealth, education, healthcare access, and other areas can undermine social stability, hinder economic progress, and exac-

Fig. 1.3 Key targets for SDG 10. 10.1. By 2030, progressively achieve and sustain income growth for the bottom 40% of the population at a rate higher than the national average. 10.2. By 2030, empower and promote the social, economic, and political inclusion of all, irrespective of age, sex, disability, race, ethnicity, origin, religion, or economic or other status. 10.3. Ensure equal opportunity and reduce inequalities of outcome, including by eliminating discriminatory laws, policies, and practices, and promoting appropriate legislation, policies, and actions in this regard. 10.4. Adopt policies, especially fiscal, wage, and social protection policies, to progressively achieve greater equality. 10.5. Improve the regulation and monitoring of global financial markets and institutions, and strengthen the implementation of such regulations. 10.6. Ensure enhanced representation and voice for developing countries in decision-making in global international economic and financial institutions in order to deliver more effective, credible, accountable, and legitimate institutions. 10.7. Facilitate orderly, safe, regular, and responsible migration and mobility of people, including through the implementation of planned and well-managed migration policies. 10.a. Implement the principle of special and differential treatment for developing countries, particularly least developed countries, in accordance with World Trade Organisation agreements. 10.b. Encourage official development assistance and financial flows, including foreign direct investment, to states where the need is greatest, particularly least developed countries, African countries, small island developing states, and landlocked developing countries, in accordance with their national plans and programmes. 10.c. By 2030, reduce transaction costs of migrant remittances to less than 3% and eliminate remittance corridors with costs higher than 5%

erbate social tensions. Below in Fig. 1.3, is a pictorial presentation of the 7 key targets.

1.4 Development of Four Countries of the IWAAC-10[1]

The focus of this book is on reducing inequality. Therefore, it is important to highlight the development trajectories of Germany, Malaysia, India, and South Africa across three key dimensions, as reflected in the Inequality-adjusted Human Development Index (IHDI). These dimensions—life expectancy, education, and income—correspond to the indicators of SDG 3 (Health), SDG 4 (Education), and SDG 8 (Decent Work), respectively. The combination of policies and programmes targeting these three dimensions can have significant impacts on reducing inequalities. The IHDI and Gender Inequality Index (GII) of Germany, India, Malaysia, and South Africa from 2010 to 2024 are presented below.

In Table 1.2, for Germany, the HDI increased from 0.926 in 2010 to 0.9.44 in 2020. However, a slight decrease was recorded in 2021. This decline, which began in 2020, is strongly correlated with the impacts of the COVID-19 pandemic. Despite the drop in overall HDI, income inequality showed a notable improvement, decreasing from 15.5 to 12.5%.

In the case of India, the HDI increased from 0.575 in 2010 to 0.663 in 2020. Over the same

[1]IWAAC-stands for inequality within and among countries.

Table 1.2 Germany

Year	HDI value	HDI change from previous year	Inequality in life expectancy	Inequality in education	Inequality in income	Gender inequality index (GII) value
2010	0.926	+ 0.003	NA	NA	NA	0.096
2015	0.938	+ 0.001	3.5%	2.6%	17.5%	0.078
2020	0.944	– 0.004	3.4%	2.7%	12.5%	0.077
2021	0.942	– 0.002	3.4%	2.7%	12.5%	0.073

Table 1.3 India

Year	HDI Value	HDI Change from previous year	Inequality in life expectancy	Inequality in education	Inequality in income	Gender Inequality index (GII) value
2010	0.575	+0.010	25.6%	42.4%	19.2%	Could not be retrieved
2015	0.929	+ 0.010	21.2%	38.7%	19.4%	0.525
2020	0.664	– 0.003	17.4%	36.9%	19.4%	0.493
2021	0.663	– 0.009	16.9%	36.9%	19.4%	0.490

Table 1.4 Malaysia

Year	HDI value	HDI change from previous year	Inequality in life expectancy	Inequality in education	Inequality in income	Gender Inequality index (GII) value
2010	0.769	+ 0.007	NA	NA	NA	0.280
2015	0.797	+ 0.005	NA	NA	NA	0.261
2020	0.806	– 0.004	NA	NA	NA	0.223
2021	0.803	– 0.003	NA	NA	NA	0.228

Table 1.5 South Africa

Year	HDI value	HDI change from previous year	Inequality in life expectancy	Inequality in education	Inequality in income	Gender inequality index (GII) value
2010	0.675	+ 0.010	NA	NA	NA	0.440
2015	0.716	+ 0.004	20.5%	14.3%	56.1%	0.427
2020	0.727	– 0.009	16.5%	17.3%	57.1%	0.408
2021	0.713	– 0.014	18.9%	17.3%	57.0%	0.405

decade, in Table 1.3 it shows that inequality in life expectancy and education decreased, while income inequality increased. The GII, however, indicates a positive trend toward reducing overall inequalities.

Malaysia showed steady progress in both HDI and GII from 2010 to 2020. However, in Table 1.4, in 2021, the HDI and GII values were recorded at 0.803 and 0.228, respectively. This reflects a decline in overall development alongside an increase in inequality, as indicated by the GII.

In South Africa as shown in Table 1.5, the HDI increased from 0.675 in 2010 to 0.727 by 2020. Furthermore, similar to Germany and India, the data indicate a reduction in overall inequalities, although a slight increase in inequality within the dimension of life expectancy was observed.

The data presented above demonstrate that social inequalities exist not only within individual countries but also among the countries under study. Variations are evident in HDI scores, inequalities in life expectancy, education, and income, as well as across gender dimensions.

Regarding HDI values, almost all countries show improvement between 2010 and 2021:

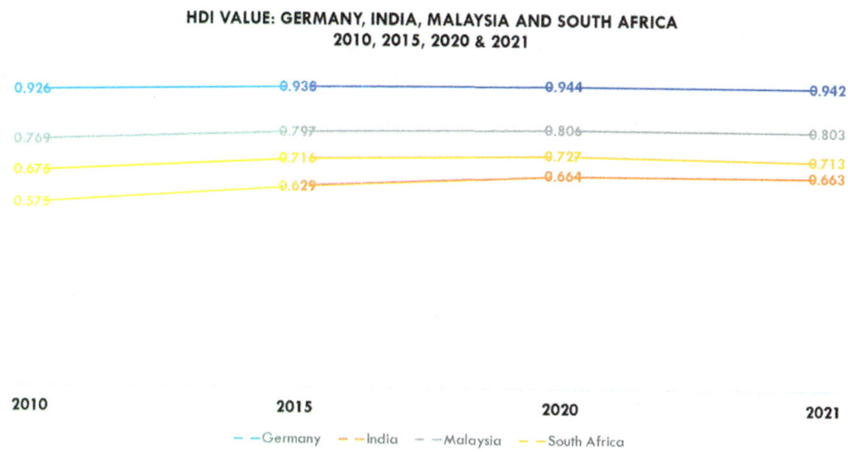

Fig. 1.4 HDI value, Germany, India, Malaysia and South Africa 2010, 2015, 2020 and 2021

Germany: 0.926 (2010) to 0.942 (2021) (very high human development).

India: 0.575 (2010) to 0.663 (2021) (medium human development).

Malaysia: 0.769 (2010) to 0.803 (2021) (high to very high human development).

South Africa: 0.675 (2010) to 0.713 (2021) (medium to high human development).

In Fig. 1.4, these developments are illustrated in the following graph:

Although steady improvement was observed over the decade, the data indicate that HDI values for all countries slightly declined in 2020–2021 (as shown by the year-on-year HDI changes). In focusing on social inequalities, the IHDI disaggregates data across three main dimensions: (i) inequality in life expectancy, (ii) inequality in education, and (iii) inequality in income. Each country has experienced inequality differently. In this context, social inequality has increased due to economic, social, and political developments at both global and national levels.

1.5 Major Challenges to SDG 10

Despite progress, significant challenges persist in achieving the targets of SDG 10. Milanovic (2016) highlights the issue of rising within-country inequality in many developed economies, driven by factors such as technological change and globalisation. Additionally, Mariotti (2019) emphasises the persistence of extreme wealth concentration, which continues to exacerbate socio-economic disparities globally.

1. **Structural Inequality**: Addressing deeply entrenched structural inequalities requires comprehensive policy interventions and reforms across multiple sectors.
2. **Data Availability**: Limited data and inadequate monitoring systems in many countries make it challenging to accurately assess progress toward reducing inequalities.
3. **Policy Coherence**: Achieving policy coherence across different sectors and levels of governance is essential but often difficult to attain due to competing interests and priorities.
4. **Global Cooperation**: Reducing inequalities within and among countries requires global cooperation, including fair trade practices, financial assistance, and technology transfer—efforts that are increasingly difficult amid geopolitical tensions and conflicting national interests.

Scholars emphasise the importance of addressing intersectionality and multiple dimensions of inequality within the framework of SDG 10. They advocate for an intersectional approach that considers factors such as race, gender, and class. UN Women highlights the gender dimension of inequality, calling for gender-responsive policies to address disparities in labour market participation and income. There have been notable efforts to address inequalities and advance SDG 10, including:

1. **Policy Reforms**: Many countries have implemented policy reforms aimed at reducing income inequality, enhancing social protection, and promoting inclusive growth.
2. **International Cooperation**: International organisations, such as the UN and the World Bank, have supported countries through technical assistance, capacity building, and funding to reduce inequalities.
3. **Civil Society Engagement**: Civil society organisations and grassroots movements play a critical role in advocating for policy changes, promoting social inclusion, and empowering marginalised communities.
4. **Private Sector Engagement**: Businesses are increasingly recognising the importance of addressing inequalities, leading to initiatives such as corporate social responsibility programmes and inclusive business models.

1.6 Implementation Strategies

There are at least two levels involved in developing strategies for SDG implementation. First, there is a need to align the SDGs with a country's National Development Plan (NDP). Second, once alignment is achieved, initiatives must be undertaken to localise the SDGs at the grassroots level. Aligning the SDGs with the NDP requires government bodies to map the SDGs to existing national policies. For example, in Malaysia, the SDGs were aligned with the Eleventh Malaysia Plan (2016–2020) and the Twelfth Malaysia Plan (2021–2025). Implementation has thus responded to local and regional needs and priorities.

In India, South Africa, and Germany, the SDGs have been implemented at different levels of government. While the SDGs are global in scope, their success ultimately depends on the ability of countries to translate them into realities within villages, cities, and regions. Localising the SDGs occurs at the grassroots level. Many SDG targets are directly linked to the responsibilities of local and regional governments, particularly in delivering basic services and promoting integrated, inclusive, and sustainable territorial development. Local spaces are therefore the critical sites for SDG delivery and implementation.

It must also be noted that the COVID-19 pandemic significantly derailed progress towards SDG implementation. Data on the IHDI and the GII from Tables 1.3, 1.4 and 1.5 for all four countries indicate a widening of inequalities between 2020 and 2021.With respect to Health (SDG 3), the pandemic strained healthcare systems worldwide, diverting resources away from other health priorities. In the area of Economic Growth and Decent Work (SDG 8), COVID-19 triggered a global economic downturn, resulting in job losses, rising income inequality, and disruptions in supply chains. The pandemic has also disproportionately affected women and girls, exacerbating existing gender inequalities (SDG 5).

Women are overrepresented in frontline healthcare and essential service roles, facing higher risks of exposure to the coronavirus (Morgan et al. 2022). Additionally, lockdown measures have led to increased incidences of domestic violence and disrupted access to sexual and reproductive health services, impacting progress towards SDG 5 on achieving gender equality and empowering all women and girls. In relation to Climate Action (SDG 13), while COVID-19 lockdowns initially resulted in temporary reductions in carbon emissions and pollution levels, the pandemic ultimately diverted attention and resources away from climate action efforts. Moreover, economic stimulus packages aimed at mitigating the pandemic's impact have, in some cases, prioritised short-term economic recovery over long-term sustainability goals.

By 2024, the Global Agenda has reached its mid-point of implementation. Countries are responsible for integrating the SDGs into their NDPs and policies, while the UN and other stakeholders support efforts to track progress, mobilise resources, and address emerging challenges. Localising the SDGs, along with robust monitoring and evaluation of progress, must form part of this process to ensure tangible benefits for beneficiaries and financial accountability.

Localising the SDGs involves adapting their content to specific local contexts, taking into account the needs of populations most affected by particular issues, the cultural norms of communities, and the commitment of stakeholders to addressing social inequalities.

Monitoring is understood as the process of gathering information to compare the actual use of project inputs and the completion of outputs against planned activities. Regular meetings with project leaders must be held to track project progress, and necessary corrective measures should be taken to address any substantial issues identified during the monitoring process.

Evaluation, by contrast, focuses on collecting information to assess the effects and overall impact of a project. Monitoring and Evaluation (M&E) is thus an integral component of any project, ensuring that inputs are effectively translated into outputs, outcomes, and impacts.

Effective M&E processes are crucial for assessing and enhancing the impact of transnational research initiatives, which are vital for addressing the challenges outlined in SDG 19. Given the global nature of these challenges, which demand collaborative and multilayered responses, robust M&E frameworks help ensure that interventions are evidence-based, contextually appropriate, and aligned with the broader goals of sustainable development.

1.6.1 SDG 10 and the Challenges for Transnational Research

Unlike traditional welfare programmes, the SDGs are holistic and multidimensional in their approach, making them more realistic in addressing the concerns of ordinary citizens. Universality means that countries should work together towards achieving the SDGs. Although local and national contexts vary, inequalities often have a global dimension, defined by historical processes and power structures. Therefore, a transnational focus—concentrating across and beyond national borders—is necessary when dealing with SDG 10 (Immler and Sakkers 2022).

Transnational issues—such as financial crises, cross-border environmental degradation, forced migration, and civic alliances for human rights, to name but a few—transcend the responsibility of any single, monolithic nation-state and highlight the need for a multilayered perspective in both research and social interventions (Schiller et al. 1992; Tedeschi et al. 2022). Consequently, there is a profound redefinition of the political field, affecting both the configuration of its context and the way politics evolves as experience, method, and practice.

Transnational research seeks to understand social problems that are global in nature—such as poverty, migration, human rights, and environmental challenges—by examining how these issues manifest differently across national contexts. This approach sheds light on diverse practices and policies, potentially identifying best practices and effective social interventions that can be adapted to different settings.

Moreover, transnational research informs policy development by providing evidence-based insights that help shape social work policies at local, national, and international levels. It thus contributes to a broader and more integrated understanding of local, national, and global developments (Negi and Furman 2010; Graßhoff and Schweppe 2012).

In summary, transnational research enhances social interventions by promoting collaboration, diversifying knowledge, and enabling the adaptation of effective strategies across different cultural and social landscapes. Global social justice has become the rallying motto of transnational social movements in world politics, where political decision-making no longer relies exclusively on nation-states, directly linking to SDG 10 (Mapp and Gabel 2022).

Research that spans multiple nations can influence policy by providing evidence that supports the effectiveness of particular interventions on a broader scale. Policymakers are often more inclined to adopt strategies that have been validated across diverse contexts. By examining how different societies address similar challenges, transnational research contributes to the development of more adaptive social systems.

With their emphasis on enhancing economic and social functioning at various levels of care, the SDGs are strongly aligned with the social work profession, reflecting social work's focus on individuals, families, groups, organisations, and communities as intervention targets to promote health and well-being. Social work operates within social relationships characterised by inequality, power, and difference, and is therefore potentially entangled in the (re) production dynamics of difference and discrimination. These factors also influence the scope of possibilities and actions available to social work professionals (Schwarzer et al. 2016; Ammann 2010; Bartley et al. 2012).

1.6.2 Significance of Transnational Academic Networks

The primary advantage of transnational academic networks is the exchange of knowledge. This exchange facilitates the flow of ideas and strategies across different contexts, leading to capacity building. Transnational research cooperation involves not only researchers from different nations collaborating on research and writing papers but also encompasses related policy frameworks, funding structures, and motivations for career development. A transnational academic network challenges local and national restrictions on academic thinking, enhancing the skills and expertise of researchers and practitioners in diverse local and national settings. This can result in new interventions informed by various national perspectives that address global challenges transcending national borders. Therefore, transnational networks can play a crucial role in challenging the national dimensions and restrictions in research, policies, and social interventions, providing collective action that can influence governmental and institutional decisions like Verband Entwicklungspolitik und Humanitäre Hilfe deutscher Nichtregierungsorganisationen e.V (VENRO 2024).

Transnational academic networks also focus on questions of positionality, location, and "situated knowledges", incorporating the role of reflexivity in the research process. The researcher's positionality raises questions related to mobility, access, and privilege, which are sometimes unveiled by adopting a transnational focus.

At the same time, transnational research networks need to be reflexive and aware of the power dynamics that exist in knowledge creation and the hegemony of what is considered "knowledge" and what is not (Freire 1993). Crossing borders is not always possible for all researchers; for example, funding might be lacking, and support may not be provided by academic or state institutions. In addition, intersectional positioning at local, national, or global levels needs to be taken into consideration when working in transnational teams or networks. Therefore, intersectionality as both a theoretical approach and a research perspective is absolutely necessary.

1.6.3 Intersectionality

Intersectionality is a tool for analysis, advocacy, and policy development that addresses multiple discriminations and helps us understand how different sets of identities impact access to rights and opportunities. While intersectionality helps us avoid the exclusion of marginalised experiences, it also urges us to address the complex, shifting realities and the constant need to devise strategies for inclusive politics (Lutz et al. 2010; Tißberger 2016; Walgenbach 2014; Crenshaw 1991; Degele and Winkler 2011; Soiland 2012).

Intersectionality aims for a more nuanced understanding by recognising and addressing the intersecting forms of discrimination and privilege that individuals experience. The theoretical framework of intersectionality was developed by Crenshaw (1989, 1991). Klicken oder tippen Sie hier, um Text einzugeben. Crenshaw highlights the specific discrimination faced by Black women in the key court case of "Emma Degraffenried versus General Motors". Her central argument is that, even when there are anti-discrimination protections based on gender and race, these do not extend to protecting Black women; instead, they primarily protect Black men and White women (Crenshaw 1991). She contends that Black women have a distinct set of experiences that cannot be adequately explained by the interaction of race and gender.

Intersectionality shapes our understanding of the multiplicity of social positionalities that individuals inhabit and within which everyone is situated. These positionalities are shaped by historical processes, such as colonialism, which created racial hierarchies and have since been adapted to changing regional, national, and local circumstances. Intersectionality enables an understanding of the complex web of hierarchical power structures in which we all live (Lutz et al. 2010; Castro Varela 2013; Castro Varela and Dhawan 2008).

According to this approach, individuals experience multiple barriers when categories such as gender intersect with others, such as race and caste. The intersection should not be understood as a simple accumulation of discriminations but must be seen as a complex interweaving of discriminations and privileges. For example, a White woman may experience sexism while simultaneously benefitting from racial privilege. Examining this complex web presents a significant challenge: the categories embedded within the research field must be identified and carefully considered, not only individually but also in terms of their intricate interrelations (Craig et al. 2008; Lee et al. 2013; Pallotta-Chiarolli and Pease 2014).

An intersectional approach seeks to address the ways in which racism, casteism, patriarchy, class oppression, religious fundamentalism, and other systems of discrimination create and reinforce inequalities that structure the relative positions of individuals. It helps us to understand and assess the impact of these structural inequalities on opportunities and access to rights. Intersectional activism, in turn, aims to address the interconnected nature of diverse hierarchical structures by advocating for policies and practices that recognise and respond to the needs of individuals who hold multiple marginalised identities.

The debate can be categorised into three perspectives on the concept of intersectionality, as introduced by McCall (2005): *inter-categorical, intra-categorical,* and *anti-categorical intersectionality*.

1. **Inter-categorical intersectionality** recognises that individuals experience multiple forms of hierarchical power structures simultaneously, due to the interaction of various social identities.
2. **Intra-categorical intersectionality** focuses on the complexity within a specific category. It acknowledges that individuals within the same group may experience discrimination or privilege differently based on their positioning in other categories.
3. **Anti-categorical intersectionality** examines the interconnectedness of social identities but advances the argument further by challenging reductive outcomes. It promotes more inclusive analyses of power and privilege, ultimately aiming to challenge hierarchical structures themselves.

McCall (2005: S.1773) notes that the three approaches "are defined principally in terms of their stance toward categories", and that they "attempt to satisfy the demand for complexity and, as a result, face the need to manage complexity, if for no other reason than to attain intelligibility". Intersectionality explains the specific nature and form of inequality and discrimination faced by certain marginalised groups and individuals, and it enables pathways to possible strategies for intervention. To enhance this understanding, it is essential to locate the historicity of social inequalities and some theoretical approaches for analysis in the case studies presented in this book.

1.7 Overview of the Book

This book is organised to demonstrate how inequalities are intersectional and located both within and among countries. It uses social inequality, intersectionality, and social justice as its core conceptual frameworks (Sect. 1.1), before illustrating the relevance of the SDGs in the transnational contexts of four countries—

Malaysia, South Africa, Germany, and India—through the lenses of social work education and social interventions (Sect. 1.2). The subsequent section (Sect. 1.3) focuses on research and social interventions undertaken in these four countries. Drawing particularly from the experience of Malaysia, Sect 1.4 highlights the relevance, necessity, and challenges of establishing concrete procedures and strategies for monitoring SDG progress from local to global levels.

Section 1.1, comprising Chaps. 2 and 3, outlines the key concepts addressed in this book and provides the conceptual framework for the subsequent chapters. It explores the relationship between social inequality, intersectionality, and the pursuit of social justice in connection to SDG 10. By examining the multiple layers of social inequality present across societies, this section highlights the effects of overlapping identities, including gender, race, ethnicity, disability, and socio-economic status. It introduces the concept of *intersectionality* as an essential perspective for understanding interrelated systems of privilege and discrimination that shape individual and group interactions. Furthermore, it explores the complex relationship between social justice and inequality, showing how structural inequalities impede the realisation of human rights and hinder efforts to build truly inclusive societies. The section emphasises that achieving substantial social justice requires addressing the root causes of inequality—namely systemic bias, unequal power distributions, and disparities in access to resources and opportunities.

Section 1.2 underscores the relevance of the SDGs. It contextualises the integration of the SDGs into social work within higher education (Chap. 4), professional practice (Chap. 5), and research (Chap. 6). This section examines the historical evolution and current practices of social work across the four countries, identifying both the opportunities and challenges associated with localising the SDGs within neoliberal contexts. It also discusses the potential of transnational social work through qualitative research with practitioners, offering insights into global collaboration and its benefits. Research is positioned as a cornerstone for generating and disseminating knowledge critical to sustainability and the reduction of inequalities. This section highlights the role of social interventions and research in addressing inequality trends and informing policies that support vulnerable communities. Furthermore, it explores intersectionality as a vital framework for tackling these issues, underlining research's critical role in advancing SDG 10 and fostering a more just and equitable world.

Section 1.3 illustrates social interventions across the four countries. These interventions are organised thematically around key issues: discriminatory practices (Chaps. 7 and 8), migration (Chaps. 9 and 10), economic inequalities and livelihoods (Chaps. 11, 12, and 13), and equal opportunity (Chaps. 14, 15, and 16).

Section 1.4 shifts the focus to evaluating the SDGs. It discusses the UN Monitoring and Evaluation (M&E) system for the SDGs across global, regional, and national levels (Chaps. 17 and 18). The section highlights both challenges and opportunities in achieving comprehensive evaluations and emphasises the urgent need for improved M&E practices to ensure the effective implementation of the SDGs, particularly in addressing the growing inequalities among countries.

In conclusion, this book offers readers—including social researchers and social work researcher-practitioners—a nuanced insight into key aspects of social inequality in the contexts of Malaysia, South Africa, Germany, and India. It calls for a transition from localised to transnational research and action to more effectively monitor and evaluate progress towards achieving SDG 10, which addresses inequalities within and among countries.

References

Ammann E (2010) Transnationale Netzwerke in der Sozialen Arbeit. http://www.soziale-arbeit.bfh.ch/uploads/tx_frppublikationen/FBS_impuls_Sept_100820_web.pdf. Accessed 26 Feb 2013

Arora-Jonsson S (2023) The sustainable development goals: a universalist promise for the future. Futures 146:103087

Bartley A, Beddoe L, Fouché C et al (2012) Transnational social workers: making the profession a transnational professional space. Int J Popul Res 2012:1–11

Bhanja R, Roychowdury K (2020) Assessing the progress of India towards sustainable development goals by 2030. J Glob Resour 06(02):81–91

Castro Varela MM (2013) Die diskursive stille unterbrechen. Kritische Migrationsforschung und politische Intervention. In: Hünersdorf B (ed) Was ist und wozu betreiben wir Kritik in der Sozialen Arbeit? Disziplinäre und interdisziplinäre Diskurse. Springer VS, Wiesbaden, pp 317–332

Castro Varela MM, Dhawan N (eds) (2008) Soziale (Un) Gerechtigkeit: Kritische Perspektiven auf Diversity, Intersektionalität und Antidiskriminierung. Lit-Verlag, Berlin

Craig G, Burchardt T, Gordon D (eds) (2008) Social justice and public policy: seeking fairness in diverse societies. Policy Press, Bristol

Crenshaw K (1989) Demarginalizing the intersection of race and sex: a black feminist critique of antidiscrimination doctrine, feminist theory and antiracist policies. Univ Chic Leg Forum 1989(1):139–167

Crenshaw K (1991) Mapping the margins: intersectionality, identity politics, and violence against women of color. Stanf Law Rev 43(6):1241–1299

Daud S, Ramli R (eds) (2022) From global to local: experiences and impacts in localising SDGs in Malaysia. Malaysian Social Science Association, Bangi

Daud S, Ramli R (eds) (2024) Sustainable development goals (SDGs) and project impact evaluation in Malaysia. Malaysian Social Science Association, Bangi

Degele N, Winkler G (2011) Intersektionalität als Beitrag zu einer gesellschaftstheoretisch informierten Ungleichheitsforschung. Berl J Soziol 21(1):69–90

Graßhoff G, Schweppe C (2012) Vom Ortsbezug sozialer Beziehungen zum transnationalen Raum—Herausforderungen für die Soziale Arbeit. Soz Passagen 4(2):171–182

Hickel J (2019) The contradiction of the sustainable development goals: growth versus ecology on a finite planet. Sustain Dev 27(5):873–884

Immler NL, Sakkers H (2022) The UN-sustainable development goals going local: learning from localising human rights. Int J Hum Rights 26(2):262–284. https://doi.org/10.1080/13642987.2021.1913411

Jossin J, Peters O (2021) Status of the SDG implementation in German municipalities. German Institute of Urban Affairs, Gütersloh

Lee MA, Smith TJ, Henry RG (2013) Power politics: advocacy to activism in social justice counseling. J Soc Action Couns Psychol 5(3):70–94

Lutz H, Vivar MTH, Supik L (eds) (2010) Fokus Intersektionalität: Bewegungen und Verortungen eines vielschichtigen Konzeptes. VS Verlag für Sozialwissenschaften, Wiesbaden

Mahlatsi K (2021) Achieving sustainable goals: challenges and prospects. Think 87(2):61–91

Mapp S, Gabel SG (2022) Moving social work from international social work to global social work through aligning SDG principles and social work education standards. Soc Work Educ 41(7):1412–1426. https://doi.org/10.1080/02615479.2022.2089647

McCall L (2005) The complexity of intersectionality. Signs 30(3):1771–1800

Morgan R, Tan H, Oveisi N et al (2022) Women healthcare workers' experiences during COVID-19 and other crises: a scoping review. Int J Nurs Stud Adv 4:100066. https://doi.org/10.1016/j.ijnsa.2022.100066

Negi NJ, Furman R (eds) (2010) Transnational social work practice. Columbia University Press, New York

Ngubane M, Pillay S (2023) An exploration of the localization of the sustainable development goals (SDGs) in municipalities: a case of south African municipalities. J Public Adm 5(1):1–10

Pallotta-Chiarolli M, Pease B (eds) (2014) The politics of recognition and social justice: transforming subjectivities and new forms of resistance. Routledge, New York

Pecquet J (2023) 10 Criticisms of the UN's sustainable development goals. https://www.theafricareport.com/322366/10-criticisms-of-the-uns-sustainable-development-goals/. Accessed 25 Apr 2025

Schiller NG, Basch L, Blanc-Szanton C (1992) Transnationalism: a new analytic framework for understanding migration. Ann N Y Acad Sci 645(1):1–24

Scholz I, Keijzer N, Richerzhage C (2016) Promoting the sustainable development goals in Germany. German Development Institute, Bonn

Schwarzer B, Kämmerer-Rütten U, Schleyer-Lindenmann A et al (eds) (2016) Transnational social work and social welfare: challenges for the social work profession. Routledge, London

Soiland T (2012) Die Verhältnisse gingen und die Kategorien kamen: Intersectionality oder vom Unbehagen an der amerikanischen Theorie. http://portal-intersektionalitaet.de/theoriebildung/ueberblickstexte/soiland/. Accessed 18 Apr 2014

Tedeschi M, Vorobeva E, Jauhiainen JS (2022) Transnationalism: current debates and new perspectives. GeoJournal 87(2):603–619

Tißberger M (2016) Hegemonial self-reflection: critical whiteness at the intersection of 'race' and gender. Paper presented at the international congress on cultural studies, University of Aveiro, 7–9 September 2016

Tiwari G, Chauhan SS, Varma R (2021) Challenges of localizing sustainable goals in small cities: research to action. IATSS Res 45:3–11

VENRO (2024) Global learning for radical change: education and the great socio-ecological and economic transformation. VENRO, Berlin

Walgenbach K (2014) Heterogenität-Intersektionalität-Diversity in der Erzichungswissenschaft. Barbara Budrich, Opladen

Part I

Social Inequalities and the Relevance of the SDGs

2 (In)Equalities, Social Inequality and Social Justice: Essential Concepts for Analysing Inequalities Within and Among Countries

Shewli Kumar and Beatrix Schwarzer

Abstract

This chapter focuses on the concepts of equality, social inequality, and social justice. Reducing inequalities requires a multifaceted approach that includes understanding and addressing the complex intersections of various forms of discrimination and privilege at different levels, specifically within institutions such as families, communities, and states (governments). Inequalities have always existed within and between countries, but they become unjust when they create barriers for certain groups of people in accessing resources necessary to lead a fulfilling life with dignity and without fear. Social justice provides a framework for promoting an equitable distribution of resources and opportunities, recognising the dynamic and ever-changing nature of social power hierarchies.

Keywords

Equality · Social inequality · Social justice · Discrimination and agency

S. Kumar
School of Social Work, Tata Institute of Social Sciences, Mumbai, India
e-mail: shewli4@tiss.ac.in

B. Schwarzer (✉)
Frankfurt University of Applied Sciences, Frankfurt, Germany
e-mail: schwarzer@fb4.fra-uas.de

2.1 Introduction

Reducing inequalities is a global challenge that takes on diverse forms at local, national, and regional levels. In this chapter, we highlight a specific interpretation of the concepts of (in)equality, social inequality, and social justice, and how we see these as related to Sustainable Development Goal 10 (SDG 10), which focuses on reducing inequalities within and among countries.

The global commitment to (i) reduce income inequality through policies that ensure fair wages, social protection systems, and progressive taxation; (ii) empower and promote the social, economic, and political inclusion of all people—irrespective of age, sex, disability, race, ethnicity, origin, religion, or economic or other status; and (iii) ensure equal opportunity and reduce outcome inequalities by eliminating discriminatory laws, policies, and practices—is not a new phenomenon. Social movements across the world, including workers' movements, women's movements, and anti- and post-colonial movements, have pursued these goals for decades (Berger and Nehring 2017; Ferree and Tripp 2006).

Therefore, efforts to address social inequalities through activism and academic conceptualisation is multifaceted (Berger and Nehring 2024; Giugni and Grasso 2019; Millward and Takhar 2019). In this book, we introduce the definitions of social inequalities and social justice that will also be used in the case study section to highlight

N. V. Roman et al. (eds.), *Social Inequalities Within and Among Countries*, Sustainable Development Goals Series, https://doi.org/10.1007/978-3-032-08741-6_2

the complexities of conducting social research and implementing interventions within the framework of SDG 10.

2.2 Equality

When discussing inequalities, the first question that arises is the question of equalities. While individuals differ and each person is unique, this alone does not mean that there are (in) equalities. Citizens of a country are generally treated equally before the law and have equal rights to access services such as education. This brings us to the first point: inequalities arise when people are treated unequally because they belong to a socially constructed group. In most cases, a binary hierarchical system of power results in one group holding privileges while the other experiences discrimination (Pease 2014; Stehr 2007). These divisions are typically based on one or more constructed commonalities, such as gender, race, caste, and other characteristics. The constructed groups are placed in a hierarchical system that assigns attributes to each group. The privileged group is typically associated with 'positive' attributes, while the discriminated group is labelled with 'negative' ones. It is important to emphasise the interdependence of these groups: efforts to reduce inequalities must address both sides and involve responsibility from both. As such, a deep understanding of power hierarchies needs must be central to all research on social inequalities. Power hierarchies refer to societal systems or structures in which certain groups possess greater power, influence, and control over others. These hierarchies can manifest at all levels of society and are often rooted in factors such as economic wealth, control over production resources, social status, race, gender, ethnicity, or other forms of privilege (Lutz et al. 2010).

Any discussion of social hierarchies is inherently a discussion about the underlying understanding of equality. Equality is sometimes defined as treating everyone the same—providing identical resources and opportunities to all individuals, regardless of their starting point or unique needs. However, this approach overlooks the fact that fairness often requires a tailored distribution of resources to address existing disparities. To effectively combat discrimination, the focus should be on substantive equality, which recognises that different groups need targeted interventions to address the challenges created by socially unjust structures (Czollek et al. 2012b; Spivak et al. 2008).

Take the right to attend school as an example: even in countries where school attendance is compulsory by law—implying equal access to education—the law alone does not ensure that all children can actually attend. This is where the principle of substantive equality becomes essential, as formal equality can turn into inequality through one or more of the following factors:

- **Structural discrimination**: This form of discrimination involves discriminatory policies, practices, and institutions, which can lead to unequal access to resources, opportunities, and outcomes (Dominelli 2002; Hooks 2000). For example, in the context of the right to education, access may be limited by factors such as inadequate transportation or a shortage of teaching staff. As a result, some children are effectively excluded from education, which can lead to reduced opportunities later in life.
- **Economic dominance**: Inequalities can be exacerbated, widening the gap between richer and poorer groups within a community or society (Chancel et al. 2022). Even when national law guarantees education for all children, the location of a school can result in unequal educational opportunities due to disparities in funding, quality of teaching support, and resources—ultimately leading to fewer opportunities for well-paid jobs in the future.
- **Marginalised communities**: These communities may face social exclusion and discrimination, which can perpetuate cycles of poverty and disadvantage (Goebel and Krause 2016; Landhäußer et al. 2010; Sobottka 2009). As a result, certain children may be excluded from schools just simply because they belong to a socially constructed group—for example, Sinti and Roma children in Germany (FRA—

Agentur der Europäischen Union für Grundrechte 2016; Rüchel and Schuch 2011), scheduled caste and scheduled tribe children in India,[1] black and coloured children in South Africa (Lam et al. 2011), or indigenous groups and refugees in Malaysia (Zamzuri et al. 2023).

- **Unequal distribution of political participation**: When political power is concentrated among the privileged, policies often serve their interests due to their disproportionate access to executive positions in political and economic systems (Bhambra 2015). In many countries, opportunities for social mobility are limited (Organisation for Economic Co-operation and Development [OECD] 2022; World Economic Forum [WEF] 2020). In the case of schooling, decisions regarding the distribution of resources—such as funding or curriculum content—are often made by a specific group of people in privileged positions. One example is representation in parliaments. In Germany, gender and ethnic groups are unequally represented in parliament, leading to the underrepresentation of their interests (Fortin-Rittberger and Kröber 2022). In India women's representation in the lower house of the parliament is below 15% and in the upper house of the parliament it is only 10% (Ghosh 2022). According to a Konrad Adenauer study, women make up just 14.86% of the lower house and 11.42% of the upper house in Malaysia, with similar patterns reflected in state legislatures (Konrad Adenauer Stiftung 2023). In contrast, South Africa has a comparatively stronger presence of women in parliament, with 46.7% representation. In Germany, only 35.3% of Bundestag members are women, placing the country 47th in global rankings according to the Inter-Parliamentary Union (2024).

[1] According to the Constitution of India, certain communities have faced historic discrimination and marginalisation, and hence, they require specific protection and affirmative action by the government. These communities have been notified as Scheduled Castes and Scheduled Tribes under Clause 1 of Articles 341 and 342 of the Constitution, respectively. https://ncst.nic.in

2.3 Social Inequality

The World Global Inequality Report (Chancel et al. 2022) highlights the stark disparities in income and wealth distribution across countries. An analysis of global data reveals that the richest segment of the population receives 52% of the global income, while the poorest receive just 8.5%. Similarly, in terms of global wealth distribution, the report shows that the poorest half of the world's population owns only 2% of the total wealth, whereas the richest 10% hold an overwhelming 76%. These figures clearly underscore deep economic inequalities both within and among countries, further exacerbated by existing social inequalities.

To explain the disparities and inequalities among countries, it is essential to draw on anti-colonial perspectives, as articulated by Shahjahan (2011), and post-colonial approaches (Said 1978; Spivak 1988). Addressing social inequality through an anti-colonial lens requires the critical examination of the historicity of colonialism, which has embedded itself in the education, bureaucracy, and judicial systems of colonised countries. Although colonised people asserted their independence and became independent nation-states, a significant degree of coloniality remained entrenched within these institutional systems. Coloniality privileged certain social groups in countries such as South Africa, Malaysia, and India, allowing them to govern their own people. This dynamic is evident in the lack of upward mobility and the exclusion of several marginalised groups from the power corridors of these systems, leading to structural discrimination and a denial of access and rights, even when these may be available in laws (de jure) and policies.

Social inequalities are contextual, societal, economic, and political, rooted in the power of some and the marginalisation of others, including individuals, communities, and nations. They are also locational; for instance, as a minority in another country or locality (from rural to urban, or region to region), marginalisation may take on different forms. Social inequalities can be understood as differences in income, resources, power, and status. These inequalities are maintained by those in power, who own and manage most of the resources in each society, country, and globally. Often, such inequalities are based on gender, race, ethnicity, disability, caste, and religion. This forms the core analytical framework of discrimination and equality embedded in SDG 10.2 and SDG 10.3, which must be addressed through social policy and social work interventions in all four countries (Germany, Malaysia, India, and South Africa).

Social inequalities operate iteratively within micro, meso, and macro systems, often creating a hierarchy based on power and disempowerment. Interpersonal inequalities and tensions perpetuate power imbalances, where certain stereotypes are deployed to 'other' individuals perceived as unfit or incapable. The concept of othering as a social process was first explained by Beauvoir (1953), who described how women were the first group deliberately 'othered' by male superiority and patriarchy. She draws on an example of a small village where all those who do not belong are regarded as 'outsiders', 'strangers', and 'suspects'.

Similarly, "to the native of a country, all who inhabit other countries are 'foreigners'; Jews are 'different' for the anti-Semite, Negroes are 'inferior' for American racists, aborigines are 'natives' for colonists, proletarians are the 'lower class' for the privileged" (Beauvoir 1953:15). Foremost among all these categories, however, are 'women'. Beauvoir (1953) emphasises that man is regarded as the absolute and autonomous being, while woman is defined only in relation to him—as the 'other'—rather than as an autonomous subject in her own right. The concept of othering was further developed by Said (1978), who explained how societies were portrayed as 'natives' and 'savages', and therefore deemed 'uncivilised'. In a similar vein, both Mohanty (1984) and Collins (1986) contributed to postcolonial feminist discourse by arguing that women are not a homogenous category. They emphasised that black women, women of colour, and women from Asian and African societies experience discrimination and exclusion in complex and distinct ways. They need a 'voice' that differs from the Western feminist understanding of patriarchy and gender inequality. In fact, the colonial narratives of these societies ensure a different kind of 'othering' and social exclusion, which follows such women wherever they may live, whether in their own societies or across the world. For example, migrant women refugees from Afghanistan and Myanmar may face deep discrimination in their everyday interactions in India, exacerbated by their migrant status (United Nations High Commissioner for Refugees [UNHCR] India and The Maldives 2022), unlike white women who arrive as tourists. Prejudice, stereotyping, violence, and deprivation act as enablers of social inequality and injustice. These ideas flow into perpetuating institutional (state, markets) inequalities through the oppressive policies and practices of organisations (educational, corporate, government systems, and others). Institutional inequalities are often unseen and hidden, perceived as the status quo, and are most challenging to identify. Structural social inequality is deeply entrenched, pervasive, and insidious, manifesting through the interaction between individuals and institutions. Discriminatory practices become normalised in everyday life, gradually embedding themselves within social systems. These inequalities are not only historically rooted but also globally sustained, making them exceptionally resistant to change.

Thus, to engage with and understand the contexts and processes of social inequalities, it is imperative to analyse what constitutes power, disempowerment, and empowerment. In her work on power, Batliwala (2019) elucidates that power is the capacity of individuals or groups to decide or influence who does what, who gets what, who sets the agenda, and who decides what. At this point in history, power cannot be claimed solely by those who have resources and control over them; it also includes the capacity to use these resources. Batliwala (2019) discusses social power and urges us to consider these four dimensions of power, explaining that it is not just control over resources but also, if we deploy gender as a category for analysis, we need to locate who has rights and who actually accesses them.

The complexity in the relationship between law, socio-economic resources, social and family norms, and power dynamics is evident in domestic violence. In Germany, laws provide protection from domestic violence (BIG Koordinierung 2021; MiMi 2018); however, statistics show an 8.5% increase in domestic violence cases reported to the police in 2022 (Bundesregierung 2023). Similarly, in India, there are stringent laws to counter violence against women, yet patriarchal gender and societal norms are deeply entrenched in women's lives. This is reflected in the rise in violence against women from 2011 to 2021, with 31.8% of women reporting domestic violence in 2021 (Navi 2023). The latest report from the National Crime Records Bureau of India (2022 pp.: xii to xix)[2] indicates a 4% increase in domestic violence cases from 2021. In Malaysia, the National Health and Morbidity Survey (Ministry of Health Malaysia, Institute for Public Health 2023) reveals that 5.9% of women have experienced emotional violence, 3.6% physical violence, and 1.6% sexual violence.

At the macro level, it is crucial to understand who sets the agenda and narratives of social power and the use of that power. For example, in a neoliberal context, it is the media that sets the agenda, determining which news is significant enough to make the front page. Agenda-setting is a form of power that is reflected at all levels, including households, national governments, international policymaking bodies, and other influential stakeholders.

To explain further, it is generally understood that the elected representatives of governments worldwide should have the power to ensure equitable distribution of resources to all, thereby reducing social inequalities. However, neoliberalism and the gradual rolling back of governments from their welfare role, along with increased privatisation and liberalisation, have unleashed a form of global inequality that serves to increase social inequalities and marginalise people and communities. For instance, in countries like India and South-East Asia, traditional communities such as fisherfolk are losing their livelihoods to commercial trawling, pollution of rivers and seas, as well as displacement due to infrastructure projects (Fabinyi et al. 2022; International Collective in Support of Fishworkers (ICSF) 2024). In this example of social inequality, we see how a top-down approach by governments can be harmful to already unequal communities and the fragile local ecology and environment.

The World Inequality Report (Chancel et al. 2022) also shows that governments have become poorer over the last 40 years. The wealth in public hands in rich countries is close to zero, with almost all of it concentrated in private hands. This has significant implications for the redistribution of wealth, as well as the responsibility and accountability of governments. Private conglomerates wield enormous economic power and will ensure that inequalities persist. The gradual abdication of state responsibility for welfare services and the redistribution of resources has led to increasing vulnerability among already excluded individuals and communities.

While social inequalities continue to plague the world order, this has led to increasing collectivisation and the coming together of individuals and groups with demands and claims for their rights. The mobilisation, collectives, and agency of individuals affected by inequality can be seen everywhere in the world. This illustrates how human beings have, and continue to have, agency to counter adversity and oppression despite deep inequality.

2.4 Social Inequality, Resistance, and Agency

Social inequality is a concern, a matter of injustice that requires attention and examination. Left unaddressed, it can lead to increased disease, hunger, poverty, conflict, disasters, and violence. This is a situation the world is already witnessing both within and among countries, whether in Sub-Saharan Africa, Asian countries like Afghanistan, or in conflicts between nations such as Israel and Palestine and Russia and Ukraine. However, these issues are being met with significant resistance, both locally and internationally. For example, large numbers of people are migrating from

[2]GOI. 2022. Crime in India 2022, Statistics Vol-1, National Crime Records Bureau, Ministry of Home Affairs, India

strife-torn areas to other parts of the world, including Europe, the United States, and the United Kingdom (UK), often in search of safety and security. This movement reflects the inherent capacity of human beings to exercise their will and resist oppression and exclusion.

This can be understood as agency, which Kabeer (1991) defines as the ability to make choices, encompassing the meaning, motivation, and purpose behind individuals' actions, as well as their sense of agency or 'power within' to change their life circumstances. Agency may take the form of negotiation, bargaining and, at times, deception and manipulation, as well as acts of subversion and resistance, alongside intangible cognitive processes of reflection and action. According to Kabeer (1999), this capability constitutes agency, which can exist at both individual and collective levels. Agency is considered positive when individuals exercise power to make life choices and pursue their own goals despite opposition and unfavourable circumstances. Conversely, agency can be negative when exercised over others through coercion, violence, and oppression.

According to Sen (1992), it is essential to understand what people are capable of being and doing, as this forms the basis of their quality of life. Quality of life is achieved through the following:

1. 'Functionings' refer to states of being and doing—what people manage to achieve in their lives. This includes actions and conditions that reflect an individual's ability 'to be' and 'to do' in ways that bring about change in their life circumstances.
2. Capability is the set of valuable functions that an individual has access to. It refers to the freedom that people must achieve the lifestyle that they wish to value. Freedom, according to Sen (1992), is the ability to act on behalf of what matters (agency), i.e. being part of institutions, democratic processes, and movements for social change.

Therefore, to understand the well-being of people, it is important to consider capabilities, i.e. the opportunities that people have, rather than just functioning, which may lead to achievements. For example, consider people who starve voluntarily by fasting versus those who starve due to a lack of access to food. Their functioning may be at the same level of deprivation, but by focusing on their capabilities, i.e. access to food, we gain a clearer picture.

Sen (1992) conducted empirical work based on his capability approach, analysing various inequalities such as gender bias, hunger, and development. Well-being serves as a useful measure of development and offers an alternative perspective, as countries with the highest income do not necessarily have the highest levels of well-being. For instance, the USA is the highest income country in the world yet does not lead when measured according to well-being (OECD 2011). Sen's theory was further expanded by Martha Nussbaum (2012) into a list of capabilities that include life, bodily integrity, bodily health, senses, imagination and thought, emotions, practical reason, affiliation, relationships with other species, play, and control over one's environment. Together, these capabilities constitute the ability to access resources, use them for well-being and social status, and develop 'agency' to act and make decisions for a better life.

2.5 Social Justice

Social justice seeks to dismantle the barriers and hierarchical power structures that sustain inequality. To do so, it brings attention to both sides of the inequality system—the discriminatory side and the privileged side (Czollek et al. 2012a). Social privilege refers to the unearned advantages individuals receive based solely on their position within a particular social group. These privileges are often invisible to those who benefit from them, as they are deeply embedded in societal norms and structures. Examples of social privilege include access to quality education, employment opportunities, healthcare, safety, and representation in media and politics. Recognising and understanding one's own privilege is essential for advancing social justice and building a more equitable society.

Social justice aims to promote a) the equitable distribution of resources, and b) equal opportunities and rights for all members of society, in order to ensure equity for every individual, regardless of their social position or identity.

Justice for equitable distribution of resources: Resources such as wealth, access to housing, jobs, and education are crucial for working towards social justice. Compared to affirmative action programmes, social justice aims to reduce and dismantle the historically rooted power systems behind social inequalities. It therefore focuses on a critical review of political agendas and power structures to find ways to dismantle them (Czollek et al. 2008, 2012a).

Justice for equal opportunities and rights: This second form of justice brings the unseen power dynamics to the forefront, such as social norms, unequal representation, and local and national values that are closely connected to privileges. This sheds light on the complex connections of discriminatory practices. Social justice is based on the idea that social differences are constructed and resists the notion of natural and unchangeable differences, which opens the socio-political space for change to reduce exclusion at a foundational level. In the concept of social justice, differences among people are seen as enriching for society and therefore always positive (Schwarzer 2014).

Social justice is therefore always critical to social power dynamics and hierarchies. It reflects on the (historical) construction of power hierarchies and structures, as well as their dynamics of change. Aiming towards greater social justice is not a single event but a dynamic and ongoing process that must adapt to a changing political and social environment, while also taking into account the local conditions of power dynamics (Czollek et al. 2012b; Herz and Johansson 2011; Lee et al. 2013).

Social justice offers a holistic approach to addressing discriminatory practices, the structural dimensions of discrimination, and underlying power dynamics. This perspective enables researchers and academics to engage with inequalities within and among countries (SDG 10) through a complex, multilayered examination of social inequalities. Academic knowledge helps to understand which discriminatory behaviours, exclusionary policies, and laws need to be addressed by socio-political and economic interventions. The core principles of social justice and bottom-up participatory approaches enable stakeholders to address inequalities and work towards greater social equity. Inequalities threaten long-term social and economic development, limiting opportunities for well-being among marginalised people and the development of self-worth. This requires governments to collaborate with social groups, communities, and civil society organisations, as outlined in SDG 17, which advocates for multi-stakeholder partnerships. While countries must take urgent steps to reverse the unequal distribution of income and resources, meaningful progress towards change and the achievement of the SDGs also requires addressing power dynamics at local, global, and transnational levels. Strengthening transnational efforts to ensure equitable resource distribution demands a firm commitment to collaborative social interventions, academic research, and policy reform, essential to addressing SDG 10.

2.6 Conclusion

This chapter explored the concepts of equality, social inequality, human agency, and social justice in depth. These concepts are essential for understanding why inequalities persist both within and among countries. The successful implementation of the objectives outlined in SDG 10 depends on the application of these concepts to first analyse the extent, nature, and type of inequalities present in different national contexts. Achieving this requires meaningful engagement with people living at the margins by all stakeholders involved in implementing the goals. Localising SDG 10 calls for the use of diverse research methods to generate key insights into the lived realities of those experiencing poverty, limited access to basic services, and social exclusion. Most significantly, it also involves understanding who makes the decisions that empower or disempower such individuals. The overarching aim of the SDGs is to promote development that is equitable, inclusive, and socially just. This chapter contributes to that aim by offering a conceptual framework for revealing previously invisible dimensions of socio-economic and political inequality across four country contexts. These concepts are further applied in the following chapter, which presents case studies from different countries.

References

Agentur der Europäischen Union für Grundrechte (2016) Bildung: Die Situation der Roma in elf EU-Mitgliedstaaten: Erhebung zur Situation der Roma–Daten kurz gefasst. https://fra.europa.eu/sites/default/files/fra_uploads/fra-2014-roma-survey-dif-education_de.pdf. Accessed 12 Nov 2023

Batliwala S (2019) All about power: understanding social power and power structures. https://creaworld.org/wp-content/uploads/2020/07/All-About-Power.pdf. Accessed 17 May 2024

Beauvoir S de (1953) The second sex. Knopf, New York

Berger S, Nehring H (eds) (2017) The history of social movements in global perspective: a survey. Palgrave Macmillan, London

Berger S, Nehring H (2024) History of social movements: palgrave studies in the history of social movements series. https://www.palgrave.com/gp/campaigns/social-movements/history-of-social-movements. Accessed 17 May 2024

Bhambra GK (2015) Citizens and others: the constitution of citizenship through exclusion. Altern Glob Local Polit 40(2):102–114

BIG Koordinierung (2021) Ihr Recht bei häuslicher Gewalt: Polizeiliche, strafrechtliche und zivilrechtliche Möglichkeiten des Schutzes. https://www.big-berlin.info/sites/default/files/downloads/330_Ihr_Recht_deutsch_2021.pdf. Accessed 17 May 2024

Bundesregierung (2023) „Häusliche Gewalt ist keine Privatsache": Zahl der Opfer von häuslicher Gewalt gestiegen. https://www.bundesregierung.de/breg-de/aktuelles/lagebild-haeusliche-gewalt-2201488. Accessed 17 May 2024

Chancel L, Piketty T, Saez E et al (2022) World inequality report 2022. https://wir2022.wid.world/www-site/uploads/2021/12/WorldInequalityReport2022_Full_Report.pdf. Accessed 17 May 2024

Collins PH (1986) Learning from the outsider within: the sociological significance of black feminist thought. Soc Probl 33(6):S14–S32

Czollek LC, Perko G, Weinbach H (2008) Radical Diversity im Zeichen von Social Justice: Philosophische Grundlagen und praktische Umsetzung von Diversity in Institutionen. In: Castro Varela MM, Dhawan N (eds) Soziale (Un) Gerechtigkeit: Kritische Perspektiven auf Diversity, Intersektionalität und Antidiskriminierung. Lit-Verl, Berlin, pp 260–276

Czollek LC, Perko G, Weinbach H (2012a) Praxishandbuch social justice und diversity: theorien, training, methoden. Übungen, Beltz Juventa, Weinheim

Czollek LC, Perko G, Weinbach H (2012b) Social justice und diversity training. Quer 18:12–20

Dominelli L (2002) Anti-oppressive social work theory and practice. Macmillan Education UK, London

Fabinyi M, Belton B, Dressler WH et al (2022) Coastal transitions: small-scale fisheries, livelihoods, and maritime zone developments in Southeast Asia. J Rural Stud 91:184–194

Ferree MM, Tripp AM (eds) (2006) Global feminism: transnational women's activism, organizing, and human rights. New York University Press, New York

Fortin-Rittberger J, Kröber C (2022) Der neu gewählte Deutsche Bundestag. Bundeszentrale für politische Bildung (9 February)

Giugni M, Grasso M (2019) Street citizens: protest politics and social movement activism in the age of globalization. Cambridge University Press, Cambridge

Goebel J, Krause P (2016) Armut in verschiedenen Bevölkerungsgruppen. http://www.bpb.de/nachschlagen/datenreport-2016/226283/armut-in-verschiedenen-bevoelkerungsgruppen. Accessed 20 Sept 2018

Herz M, Johansson T (2011) Critical social work–considerations and suggestions. Crit Soc Work 12(1):28–45

Hooks b (2000) Feminist theory: from margin to center. South End Press, Cambridge

International Collective in Support of Fishworkers (ICSF) (2024) India: world fisheries day 2023: resistance and struggles of fisherfolks. https://www.icsf.net/newss/india-world-fisheries-day-2023-resistance-and-struggles-of-fisherfolks/. Accessed 17 May 2024

Inter-Parliamentary Union (2024) Women in national parliaments. http://archive.ipu.org/wmn-e/classif.htm. Accessed 18 Mar 2025

Kabeer N (1991) Gender, development, and training: raising awareness in the planning process. Dev Pract 1(3):185–195

Kabeer N (1999) Resources, agency, achievements: reflections on the measurement of women's empowerment. Dev Change 30(3):435–464

Konrad Adenauer Stiftung (2023) Challenges for women in political parties in Malaysia and acceleration strategies to leadership in politics. https://www.kas.de/documents/272088/23299843/E-Book+2023-02-08+Challenges+for+Women+in+Political+Parties+in+Malaysia.pdf/f9e65137-c8f2-0f59-7761-4b9d4dcd9379?version=1.1&t=1686197816307. Accessed 18 Mar 2025

Lam D, Ardington C, Leibbrandt M (2011) Schooling as a lottery: racial differences in school advancement in urban South Africa. J Dev Econ 95(2):121–136

Landhäußer S, Kessl F, Klein A (2010) Armut und Prekarisierung von Adressat_innen Sozialer Arbeit. In: Thole W (ed) Grundriss soziale Arbeit: Ein einführendes Handbuch. VS Verlag für Sozialwissenschaften, Wiesbaden, pp 541–548

Lee MA, Jorgensen Smith T, Henry RG (2013) Power politics: advocacy to activism in social justice counseling. J Soc Action Couns Psychol 5(3):70–94
Lutz H, Herrera Vivar MT, Supik L (eds) (2010) Fokus Intersektionalität: Bewegungen und Verortungen eines vielschichtigen Konzeptes. VS Verlag für Sozialwissenschaften, Wiesbaden
Millward P, Takhar S (2019) Social movements, collective action and activism. Sociology 53(3):NP1–NP12
MiMi (2018) Violence protection for women in Germany. https://www.mimi-gegen-gewalt.de/wp-content/uploads/2019/03/RatgeberGender_ENG.pdf. Accessed 17 May 2024
Ministry of Health Malaysia, Institute for Public Health (2023) National health and morbidity survey: maternal and child health. https://iku.gov.my/images/nhms-2022/4a.-infographic_nhms-mch_english_19.05.2023.pdf?_t=1684481335. Accessed 18 Mar 2025
Mohanty CT (1984) Under western eyes: feminist scholarship and colonial discourses. Boundary 2 12(3):333–358
National Crime Records Bureau (2022) Crime in India. https://www.ncrb.gov.in/uploads/nationalcrimerecordsbureau/custom/1701607577CrimeinIndia2022Book1.pdf. Accessed 24 Sep 2024
Navi F (2023) The case for a feminist approach to gender-based violence policymaking in India
Nussbaum MC (2012) Philosophical interventions: reviews, 1986–2011. Oxford University Press, Oxford
Organisation for Economic Co-operation and Development (OECD) (2022) Understanding social mobility: promote equal opportunity. https://www.oecd.org/stories/social-mobility/. Accessed 17 May 2024
Pease B (2014) Transforming privileged subjectivities: toward a pedagogy of the oppressor. In: Pallotta-Chiarolli M, Pease B (eds) The politics of recognition and social justice: transforming subjectivities and new forms of resistance. Routledge, New York, pp 159–172
Rüchel U, Schuch J (2011) Bildungswege deutscher Sinti und Roma. https://mediendienst-integration.de/fileadmin/Dateien/2011_Strauss_Studie_Sinti_Bildung.pdf. Accessed 5 June 2021
Said EW (1978) Orientalism. Vintage Books, New York
Schwarzer B (2014) Ansätze für eine diversity-sensible Soziale Arbeit. In: Bretländer B, Köttig M, Kunz T (eds) Vielfalt und Differenz in der Sozialen Arbeit: Perspektiven auf Inklusion. Kohlhammer, Stuttgart, pp 195–205
Sen A (1992) Inequality reexamined. Oxford University Press, Oxford
Shahjahan RA (2011) Decolonizing the evidence-based education and policy movement: revealing the colonial vestiges in educational policy, research, and neoliberal reform. J Educ Policy 26(2):181–206
Sobottka EA (2009) Armut und Armutsfolgen in den Ländern der peripheren Moderne. In: Wagner L, Lutz R (eds) Internationale Perspektiven sozialer Arbeit: Dimensionen—Themen—Organisationen. VS Verlag für Sozialwissenschaften, Wiesbaden, pp 137–153
Spivak GC (1988) Can the subaltern speak? In: Nelson C, Grossberg L (eds) Marxism and the interpretation of culture. University of Illinois Press, Urbana, IL, pp 271–313
Spivak GC, Steyerl H, Joskowicz A et al (2008) Can the subaltern speak? Postkolonialität und subalterne Artikulation. Turia + Kant, Wien
Stehr J (2007) Normierungs- und Normalisierungsschübe-Zur Aktualität des Foucaultschen Disziplinbegriffes. In: Anhorn R, Bettinger F, Stehr J (eds) Foucaults Machtanalytik und Soziale Arbeit: Eine kritische Einführung und Bestandsaufnahme. VS Verlag für Sozialwissenschaften, Wiesbaden, pp 29–40
United Nations High Commissioner for Refugees (UNHCR) India and The Maldives (2022) A place for all refugees under India's welfare umbrella. UNHCR India, 21 November
World Economic Forum (WEF) (2020) The global social mobility report 2020: Equality, opportunity and a new economic imperative. https://www3.weforum.org/docs/Global_Social_Mobility_Report.pdf. Accessed 17 May 2024
Zamzuri NH, Hanafiah MH, Suzila TI et al (2023) The connection between missing school and indigenous community: challenges and complexities. AlterNative 19(3):725–730

3 Social Work, Social Inequalities, and the Sustainable Development Goals (SDGs)—A Four Country Account

Lea Teresa Schaefer, Anne Chmela, Meera Suresh Babu, Nele Tyburski, Sophie Weihs, and Dagmar Oberlies

Abstract

This overview presents four country profiles: Germany, India, Malaysia, and South Africa. These countries took part in a DAAD-funded SDGs-partnership programme to explore inequalities within and among countries (IWAAC-10). The country profiles are structured using a toolkit designed to map social services, focusing on the socio-economic context, laws and policies, education and training, and the social sector workforce. In the final section, we discuss the similarities and differences among the four countries and draw conclusions from a transnational perspective.

Keywords

Social services · Social work education · Social inequalities · Social protection policy · Sustainable development goals · Decolonizing social work

3.1 Introduction

The 2030 Agenda and its interrelated Sustainable Development Goals (SDGs) provide a comprehensive framework to 'transform our world' (A/RES/70/1). The active role of social work, social services, and community engagement is essential to realising this vision (Jayasooria 2016; International Federation of Social Workers [IFSW] 2017; Seth 2018; Nikku and Pulla 2014; Kaltenborn et al. 2020; Pandey and Kumar 2018).

Of the 248 indicators associated with the SDGs, more than one-quarter are directly relevant to social work and social services. One of these goals—and the central focus of this book—is SDG 10: Reduced Inequalities. This goal addresses key issues such as poverty, social inclusion, equal opportunity, non-discrimination, social protection policies, and safe and responsible migration.

While the SDG monitoring system relies heavily on quantitative indicators, statistical data, and randomised trials, social work and

L. T. Schaefer (✉) · A. Chmela · N. Tyburski · S. Weihs · D. Oberlies
Department of Social Work and Health, Frankfurt University of Applied Sciences, Frankfurt am Main, Germany
e-mail: lt.schaefer@alumni.maastrichtuniversity.nl

A. Chmela
e-mail: anne@chmela.de

N. Tyburski
e-mail: nele.tyburski@gmx.de

S. Weihs
e-mail: sophie.weihs@t-online.de

D. Oberlies
e-mail: Dagmar.Oberlies@t-online.de

M. S. Babu
School of Social Work, Tata Institute of Social Sciences, Mumbai, Maharashtra, India
e-mail: meerasuresh1994@gmail.com

N. V. Roman et al. (eds.), *Social Inequalities Within and Among Countries*, Sustainable Development Goals Series, https://doi.org/10.1007/978-3-032-08741-6_3

applied social research must engage with complex social problems embedded in cultural contexts and multilayered social interventions. This complexity calls for:

- A multidisciplinary approach that draws from various subfields, including political science, development studies, sociology, developmental psychology and family studies, gender studies, social work, criminology, and law (United Nations Educational, Scientific and Cultural Organisation [UNESCO] 2016).
- An education-research-practice-policy nexus that addresses discriminatory practices and fosters a rights-based approach. It should also promote inclusive, research-based interventions and strengthen community participation and ownership.
- A transnational mode of knowledge production that both decolonises and localises knowledge and methodology.

(Higher) education plays a critical role: it produces 'skilled practitioners', educates global citizens, mainstreams sustainable development, and introduces innovative degree programmes.

In the following, the social service landscape in Germany, India, Malaysia, and South Africa will be explored sequentially. The country profiles are structured using a toolkit designed to map social services, focusing on the socio-economic context, laws and policies, education and training, and the social sector workforce. In the final section, we discuss the similarities and differences among the four countries and draw conclusions from a transnational perspective.

3.2 Methodology

For this overview, a toolkit for mapping social services—particularly the literature review matrix—was adopted for each country (Global Social Service Workforce Alliance [GSSWA] 2019). To describe social inequalities in a consistent manner, we drew on reports and dashboards that compare SDG-related indicators across countries (Sachs et al. 2024; World Bank 2024b; United Nations [UN] 2024). The country profiles focus on targets related to reducing inequalities, such as income disparities, social inclusion, non-discrimination, migration, and integration. However, not all SDG-related indicators and data were available for each of the four countries, and in many cases, the data relied on uncertain estimates (Ghatak and Rishabh 2024). To compare migration policies, we consulted the Migration Integration Policy Index (MIPEX) (Solano and Huddleston 2020a).

The description of social services draws on a range of resources and literature. In the first step, we searched two databases—ResearchGate and the GSSWA database—using keywords such as 'social work' and 'social service' in combination with specific countries or regions. Additionally, we conducted online searches using terms like 'social work/social services' and 'bill' or 'act'. In the second step of the process, consultation was undertaken with partners, who were invited to make recommendations regarding relevant materials. All accessible documents published after 2015 were included in the review matrix and compiled into a country profile. The reviewed materials encompassed laws and policies, academic articles, and grey literature such as curricula, proposals, reports, and toolkits.

3.3 Country Profiles

3.3.1 Germany

Social Inequalities

According to the 2024 Sustainable Development Report, 11% of the population in Germany is classified as 'poor', as their disposable income falls below half the national median (Sachs et al. 2024:213). The richest 10% have as much as the poorest 40% (Palma ratio: 1.1). Nevertheless, income inequality remains moderate (Gini coefficient: 31.7). Groups identified as being at risk of income poverty include the unemployed, low-skilled workers, single mothers, large families, non-German residents, senior citizens, and even students—with women disproportionately affected (Bundesministerium für Arbeit und Soziales 2021).

Disaggregated data by gender, age, disability, and other social categories remain limited (Sachs et al. 2024:83). Available figures include a poverty rate of 14% among the elderly and 8% for youth not in employment, education, or training (NEET). Gender-related indicators reveal fewer years of education, lower labour force participation, and reduced wages for women. However, data on gender-based violence (GBV) remain unavailable—even on UN Women's SDG Indicator Dashboard. Persons with disabilities are not mentioned in the available datasets.

Germany has signed and ratified all major human rights instruments aimed at protecting individuals from discrimination and has enacted the Equal Treatment Act (AGG). Nevertheless, data from the national statistical office indicate that one in twelve users of public services and one in ten workers report experiencing discrimination (DeSTATIS 2021). Moreover, a survey conducted by the Federal Anti-discrimination Agency found that one in three respondents had experienced discrimination (Antidiskriminierungsstelle des Bundes 2022:13).

According to the IOM World Migration Report of 2024, nearly one-fifth of Germany's population consists of migrants (McAuliffe and Oucho 2024), including 2.6 million refugees (World Bank 2024a). However, Germany's migration policies are assessed as only 'halfway favourable', scoring 58 out of 100 across eight key policy areas (Solano and Huddleston 2020b).

Policies and Regulations

Germany is defined as a 'social state' (Art. 20 Basic Law), implying the guarantee of a level of subsistence, social protection, and equal opportunities (Bundestag 2014). The Social Codes (SGB) cover a wide range of areas, including social insurance, social assistance, child and youth welfare, rehabilitation and participation, as well as various other social benefits. Courts have ruled that these entitlements impose a duty on the state to maintain and fund the corresponding social services (cf. BVerwG 3 C 26.02). Accordingly, the codes also address social services such as youth assistance, support for adults facing social difficulties, and care for individuals in need. The tasks and roles of social services are explicitly defined within the criminal justice system and may also be specified by state directives or broadly recognised standards, such as those for victims of violence or refugees.

Social work in Germany is a 'regulated profession'. Each 'Länder' (state) has enacted legislation governing the recognition of social workers, and a licence is required to practice. The minimum requirements include a bachelor's degree after six semesters of study (180 ECTS) in accordance with the Qualifications Framework for Social Work, along with at least 100 days of practical experience under the guidance of a licenced social worker and the supervision of university teaching staff (Schäfer and Bartosch 2016).

Education and Training

German universities offer 285 undergraduate and 183 graduate programmes in Social Work (Hochschulkompass 2025). In terms of enrolment rates, Social Work ranks eighth, with over 120,000 students—predominantly undergraduates at Universities of Applied Sciences (84%)—and only 1% pursuing doctoral studies (DeSTATIS 2024a). A survey further revealed that Social Work is 'conspicuously' gender-segregated with a strong *career* orientation (Pulver and Matti 2021).

The German Qualification Framework for Social Work (Schäfer and Bartosch 2016) defines core competencies across several areas: knowledge and understanding, analysis and evaluation, planning and conception, implementation and assessment, as well as professional and personal skills and attitudes. Based on this framework, universities design and develop degree programmes for accreditation.

One such programme is Social Work: Transnational at Frankfurt University of Applied Sciences (UAS). It recognises that social work faces new challenges arising from global developments such as climate change, conflict, and migration. The syllabus integrates the humanities, social sciences and law, socio-economic inequalities, intersectional discrimination, and interdisciplinary aspects of migration. A key feature of the programme is the requirement of two mandatory stages abroad: one Erasmus study semester and one field placement outside of Europe. Combined with an internship in Germany, primarily in refugee centres and migration services, these experiences fulfil the 100-day practical requirement for the licence in a four-year programme (Chmela 2024). Further study opportunities at the master's level include Psychosocial Counselling and Law, Research in Social Work, and Diversity and Inclusion, as well as a PhD in Social Work (Hochschulkompass 2025).

Social Service Workforce

Every year, nearly 20,000 students graduate in Social Work and join a workforce of nearly half a million, composed predominantly of women and often in part-time roles (Bundesamt für Arbeit 2024). Only around 7000 social workers register as unemployed annually, despite there being four times as many job openings. Nevertheless, even though there is a significant increase in personnel, substantial shortages of qualified social work professionals persist (Statista 2024).

Twenty percent of the social service workforce are public servants, a similar proportion work in various supervised centres, and ten percent are employed in education (Bundesamt für Arbeit 2024). In child and youth services, the majority are employed by private entities, including homes (100,000), family counselling (55,000), or school-related services (43,000). Social pedagogues can also enter childcare, which has more than 700,000 caregivers (DeSTATIS 2024b).

Notable information is the ratio of social workers to clients:

- One full-time probation officer supervises between 30 and 80 probationers—youth assistance is limited to 45 cases (Landgericht Frankfurt am Main 2022).
- A network of disability workshops employs 70,000 professionals for 310,000 clients, primarily with mental (75%) or physical limitations (3%) (Rehadat 2022).
- Women's shelters and counselling centres for violence against women employ at least one social worker (Deutscher Bundestag 2012:54, 83).
- In refugee facilities, one social worker handles 60 to 80 clients, while an integration advisor may be responsible for up to 250 clients (Deutsche Presse-Agentur [DPA] 2023).

Social work in Germany is organised in various ways. Larger welfare organisations represent social service providers at both the federal and state levels. Professional associations, the largest of which is the German Association of Social Workers (DBSH), along with numerous smaller sectoral or ideological associations, represent the interests of their members. Working groups are formed for programmatic discussions and advocacy.

3.3.2 India

Social Inequalities

India is currently (2024) the world's most populous country, with 60% of its population residing in rural areas (Ghatak and Rishabh 2024). In recent years, India has made significant progress in reducing extreme poverty (World Bank 2024b). Nevertheless, about one-fifth of the population remains below the national poverty line (Sachs et al. 2024). Half of the population earns less than a lower-middle income (Gini coefficient: 34.2), while 10% of the population controls 77% of the country's wealth, which is three times the wealth of the poorest 40% (Palma: 3.1). In addition to monetary indices, India also uses a Multidimensional Poverty Index (MPI), which simultaneously assesses deprivations in health, education, and standard of living.

Farm labourers in rural areas, casual workers in urban centres, and tribal communities, including Dalits and Muslims, continue to be among the most impoverished groups (Rai and Mishra 2024: 1840–1841), with one third of the youth classified as NEET and women remaining trapped in precarious, often unpaid labour (World Bank 2024b).

India is a signatory to all major human rights treaties (Office of the High Commissioner for Human Rights [OHCHR] 2024). Despite numerous legal measures, discrimination continues to contribute to significant social inequality. Oxfam found that gender and caste-based discrimination affects access to the labour market; however, this is less pronounced for Muslims, who possess special professional skills acquired over generations (Oxfam India 2022). The prevalence of partner violence remains high (UN Women 2018).

The proportion of migrants living in India is below 0.5% (McAuliffe and Oucho 2024). Approximately 46,000 refugees and asylum seekers are registered with the United Nations High Commissioner for Refugees (UNHCR), primarily from Myanmar and Afghanistan (World Bank 2024a). In terms of integration, India still faces significant challenges (Solano and Huddleston 2020c).

Policies and Regulations

As per Article 38 of the Indian Constitution, the state is mandated 'to promote the welfare of the people'. Ideal policies should aim for adequate means of livelihood, ownership that serves the common good, no concentration of wealth, equal pay, and the protection of children (Art. 39). As these are 'Directive Principles of State Policy', social rights are contingent upon government action (Khosla 2010).

Social work in India is constitutionally guided by principles of justice, equality, and human dignity. The role of social workers is specifically mentioned in certain areas, such as the Juvenile Justice Act (2015) and the Mental Health Act (2017). These laws establish the participation of social workers in the Child Welfare Committee, the Juvenile Justice Board, and in multidisciplinary treatment and rehabilitation (United Nations Children's Fund [UNICEF] 2018: 43; Sahu 2017).

Nevertheless, social work in India continues to face challenges in gaining recognition as a distinct profession (Dash 2017). Draft bills to establish a National Council of Professional Social Work—with the authority to create rules, regulations, and set standards for education—have been tabled repeatedly since 1993 (Singh and Saumya 2021: 15; Actovin and Prasad 2023). In 2020, yet another draft was circulated by the government (Mathew 2020; National Council of Social Work [NCSW] 2020).

Education and Training

Social work education began in India in 1936 when the Tata Graduate School of Social Work (now known as TISS) introduced a two-year training programme in Social Work (Das et al. 2022:1; Desai et al. 2020; Harini et al. 2023). A working group on social work education estimated in 2021 the existence of over 500 social work education institutions in India (Working Group of Professional Social Workers 2021: 2). This expansion has led to growing demands for national standards and accreditation (Working Group of Professional Social Workers 2021; Actovin and Prasad 2023).

In its initial phase, social work education in India relied heavily on American curricula and literature, with limited emphasis on community development (Adaikalam 2014; Prasad et al. 2024). From there, it evolved through several phases (Singh and Saumya 2021; Singh 2021). The last official review report was published in 1980, and the most recent model curriculum dates back to 2001 (Working Group of Professional Social Workers 2021; Singh 2021). Finally, a voluntary initiative published a Curricular Framework for Social Work Education (National Curriculum Resource Group 2022). The framework outlines key subject areas, with a primary focus on community engagement. Its list of 'core areas' brings together 17 topics, including: National/International Social Welfare/Social Justice Policies, Legislation and Programmes (for socially excluded groups such as women, Dalits, tribals, landless and small farmers, the organised and unorganised labour, children, adolescents, older persons, persons with disabilities, religious, ethnic/racial, and sexual minorities, etc.).

TISS, as the oldest social work institution in India, initially focused on postgraduate studies (Singh 2021: 1061). Today, the School of Social Work (Mumbai) comprises nine centres, each offering a master's degree, along with one joint PhD programme and numerous short-term certificate and diploma courses (TISS 2024a). The master's programmes include Community Organisation and Development Practice, Dalit and Tribal/Disability Studies and Action, and Women-Centred Social Work Practice (TISS 2024b).

Since 2004, TISS has offered an undergraduate programme in Social Work at its rural campus in Tuljapur. The academic report for 2021/2022 announced the launch of an additional bachelor's programme on the Mumbai campus and committed to aligning all undergraduate and postgraduate programmes with the SDG framework (TISS 2023). In accordance with the New Education Policy, the new and revised programmes commenced almost simultaneously. Notably, both programmes share many courses drawn from a universal catalogue ('baskets') of humanities, social sciences, and social work (Singh 2021: 1061). Certain courses at the rural campus have been designed to enhance students' proficiency in both English and Marathi, in addition to providing an introduction to world literature. The rural programme is evidently more connected to nature, while the urban programme dedicates more hours to research and slightly less to field exposure beyond research. Due to its larger campus, students in Mumbai can choose electives from across the institution. The rural programme offers a certificate after one year, a diploma after two, and a BA (Hons) after three years. Graduation with a research degree is possible after four years, with one disciplinary major (Social Work) and two minors: Rural Development and Micro Practices (TISS 2024C).

Social Work Education in India equips students with the skills to understand and respect diverse beliefs and their influence on individual and social behaviour (Vareed 2023). The TISS Bachelor of Social Work Curriculum Draft (2023) emphasises intersectionality and the ability to work effectively at the intersections of caste, gender, and other social categories.

Social Service Workforce

The workforce in India consists of nearly half a million social workers (Working Group of Professional Social Workers 2021). They are employed by civil society organisations, government bodies, the private sector, and faith-based organisations. Since no licencing or registration is required, little additional information is available.

Furthermore, India has four national professional associations for social work, several state-level organisations, and the Indian Network of Professional Social Work. On international level, Indian representatives have been very active in international associations (Working Group of Professional Social Workers 2021).

3.3.3 Malaysia

Social Inequalities

Since gaining independence in 1957, Malaysia has transformed from an agricultural economy into a diversified industrial and service-based economy. However, due to the impact of COVID-19, the country has not yet returned to pre-pandemic poverty levels; in 2024, 6% of the population lives below the national poverty line (World Bank 2024b). While urban poverty has increased, rural poverty rates have declined—presumably due to subsistence farming. However, economic disparities remain high (Sachs et al. 2024). Poverty and inequality in Malaysia have an ethnic and regional dimension (Ravallion 2020; World Bank 2024b). The Palma Ratio for 2021 registered at 1.92, indicating that the wealthiest 10% of the population hold assets that are twice the value of those held by the poorest 40%. A group commonly referred to as the 'B40' in Malaysia (Sachs et al. 2024).

The Malaysian population is divided into several ethnic groups. According to official statistics,[1] about 60% are native Malays or indigenous

[1] jdn.gov.my

people, 21% are of Chinese descent, and another 6% are identified as 'Indian', although most were born in Malaysia. Article 153 of the Malaysian Constitution grants a 'special position' to Malays and natives from Sabah and Sarawak, a policy introduced following the ethnic clashes of 1969. It ensures quotas for entry into public education and the civil service. Unemployment is overall low, with fewer women participating than men (Sachs et al. 2024).

Data on GBV (SDG 5.2) are not available. However, Malaysia established a National Committee on Domestic Violence in 2019 and amended legislation to eliminate violence against women (Economic Planning Unit 2021: 195, 221). Malaysia also ratified the international Convention on the Elimination of All Forms of Discrimination Against Women (CEDAW) (Office of the High Commissioner for Human Rights [OHCHR] 2024).

The Malaysian Department of Statistics estimated three million migrants, both documented and undocumented, are living in Malaysia as of July 2023. Most originate from neighbouring Asian countries and many work in manufacturing, plantations, agriculture, hospitality, and domestic work. Additionally, the UNHCR has registered 180,000 refugees, primarily Rohingya from Myanmar (International Organisation for Migration [IOM] 2024). The government of Malaysia views the high numbers of illegal immigrants as a significant challenge concerning crime and public safety (Economic Planning Unit 2021:154). An evaluation of integration policies is not available for Malaysia.

Policies and Regulations

The Malaysian Constitution grants every person fundamental liberties, as well as protection against forced labour and, in general, against discrimination (unless authorised by the constitution). The federal government has jurisdiction to pass social welfare legislation, including the Employees' Social Security Act (1969) and the Minimum Wages Order (1/2023). Nevertheless, the family is viewed as the primary care provider (Koutronas 2020:23).

Malaysia has signed key international treaties, including the Convention on the Elimination of All Forms of Discrimination Against Women (CEDAW), the Convention on the Rights of Persons with Disabilities (CRPD), and the Convention on the Rights of the Child (CRC). These commitments have led to the enactment of the Domestic Violence Act (1994), the Child Act (2001), and the Persons with Disabilities Act (2008).

A Social Work Bill has been under discussion since 2010 and is currently (2024) under review by the Ministry of Women, Family and Community Development (Hatta et al. 2014:140; UNICEF and Malaysian Association of Social Workers [MASW] 2023; Hua and Shaffie 2016; Hua 2023). The bill aims to align social work standards with the global benchmarks of the profession (Amin et al. 2022:2364). In the absence of such legislation, social work in Malaysia continues to lack regulations for the recruitment, registration, and licencing of trained social workers (Fuziah and Teoh 2017: 120).

Education and Training

Western theories, models, and literature continue to dominate social work education in Malaysia, largely due to the country's colonial legacy (Ling 2002:85; Mohammed et al. 2020; Das et al. 2022). Today, social work education is offered at eight public and four private institutions, covering programmes from diploma to PhD level (UNICEF and MASW 2023). The programmes are accredited by the Malaysian Qualification Agency (MQA), which is the general accreditation body for all higher education programmes (Amin et al. 2019:48; Hua 2023; Amin et al. 2022).

The undergraduate programme at Universiti Kebangsaan Malaysia (UKM) addresses issues such as rapid urbanisation, changes in family structure, and decreasing social support systems. Despite calls to localise social work (Das et al. 2022), many topics reflect the universal curriculum, including poverty, unemployment, unequal opportunities, racism, crime, domestic violence, substance abuse, and mental illness. The programme also emphasises a core set of skills: social, communication, and management skills, along with professional competencies and crisis intervention. Beginning at the undergraduate level, students conduct research to address problems affecting marginalised groups and to facilitate change at the individual, group, and community levels. The three-year programme includes a six-week placement in an organisation that provides social services. Local providers suggest aligning undergraduate programmes more closely with the SDGs (Jayasooria 2016:28; Azizul and Amin 2019). UKM also offers an MPhil to enable students to study for a PhD in Social Work.

Social Service Workforce

In 2022, the Social Welfare Department employed almost 4000 professional social workers, working across welfare and probation offices, as well as anti-trafficking units (UNICEF and MASW 2023). Other important areas of practice include mental health and substance abuse treatment, medical social work, family and child services, and ageing care (MASW 2024).

The social service workforce in Malaysia includes both 'professional' social workers and 'functional alternatives' (Hatta et al. 2014). 'Functional alternatives' play a crucial role, particularly given the shortage of trained social workers. In 2020, the Department of Social Welfare announced a five-year plan to enhance its services, professionalise its workforce, and better address societal needs (Ministry of Health Malaysia 2020; Jabatan Kebajikan Masyarakat Malaysia [JKM] 2020). While working with NGOs offers job opportunities for social workers, it is not yet widely established as a viable career path (Amin et al. 2022:2367; Parker et al. 2017). Trained social workers in Malaysia are organised through the MASW, which is a member of the International Federation of Social Workers.

3.3.4 South Africa

Social Inequalities

South Africa is one of the most unequal countries in the world (Gini coefficient: 63). The latest available statistics (2014/2015) show that more than half of the population lived below the national poverty line, about 25% experienced food poverty, and close to one-third were unemployed (World Bank 2024b). The richest 10% accumulated almost seven times more wealth than the poorest 40% (Palma Ratio: 6.9).

In terms of education and labour market participation, South Africa appears relatively gender egalitarian (Sachs et al. 2024). Nevertheless, 13% of women aged between 15 and 49 have experienced partner violence (UN Women 2018). Citizens consider it a 'common occurrence' (Afrobarometer 2023) and the most significant women's rights, social, and health issue (Leburu-Masigo 2020:177).

The country's history is marked by colonialism and a system of institutional racial segregation (apartheid) (Sakala et al. 2022:7). Section 9 of the Constitution (Republic of South Africa [RSA] 1996) explicitly condemns discrimination and guarantees protection towards equality. Post-Apartheid South Africa has ratified several UN treaties addressing discrimination (OHCHR 2024). However, discrimination persists: in a slightly older quality of life survey, one in ten persons of colour reported experiencing discrimination (despite a strong sense of belonging), while others perceived that young men and black Africans were 'excluded'—though not women (Global Centre for Pluralism 2023:48).

The MIPEX asserts that South Africa's integration policy is regarded as unwelcoming (Solano and Huddleston 2020d). After a sudden influx, it withdrew status and protection from refugees in 2017, and in 2020 it started deporting migrants for taking part in a demonstration against xenophobia (Global Centre for Pluralism 2023:17). Currently, 67,145 refugees and asylum seekers are registered in South Africa, representing approximately 0.1% of the total population (World Bank 2024a).

Policies and Regulations

Section 27 of the South African Constitution guarantees the right to health care, food, water, and social assistance, and obliges the state to 'take reasonable legislative and other measures, within its available resources, to achieve the progressive realisation of each of these rights'. The principle of *ubuntu*—defined as 'an individual's humanity expressed through his relationship with others'—was introduced in a postscript to the interim Constitution (RSA 1996; Mpedi 2017:37).

Rooted in the constitution, South Africa's laws on social insurance and assistance form the foundation of its social welfare system. As such, the country adheres to the minimum standards established by the International Labour Organisation (ILO), including provisions for medical care and insurance coverage for illness, unemployment, retirement, workplace injuries, family support, maternity leave, disability, and survivor benefits (Mpedi 2017:41; Gray and Lombard 2022). The South African Social Assistance Act (RSA 2004) provides for Child Support Grants (CSGs), care for persons with disabilities, support for older persons, and assistance for war veterans (Mpedi 2017:42; Chiwara 2024; Sakala et al. 2022). Tax-funded grants contribute to the availability of necessary social services, as envisioned in the 1997 White Paper for Social Welfare (Noyoo et al. 2021:208; Lombard and Twikirize 2014; Sakala et al. 2022).

The social work profession in South Africa is regulated by the Social Service Professions Act (RSA 1978) and governed by the South African Council for Social Service Professions (SACSSP). The SACSSP establishes standards for professional practice and training, and mandates the registration of both students and practitioners. This registration serves as a licence to practice (SACSSP 2024) and must be renewed annually (Lombard and Twikirize 2014:318; Lombard et al. 2003).

Education and Training

The principles for social work education, in line with the National Qualifications Framework (NQF), are defined by the South African Qualifications Authority (SAQA) (Council on Higher Education [CHE] 2015). These principles require graduates to address critical social challenges such as unemployment, child abuse, domestic violence, crime, substance abuse, and the impact of HIV/AIDS. Graduates are expected to tackle the root causes of poverty and inequality, empower individuals and communities, uphold social work values and human rights, and work effectively in diverse, multidisciplinary teams (SAQA n.d.). According to Educations.com, there are 17 social work programmes in South Africa, with focal areas including Community Service (2), Counselling (6), Human Services (1), Social Care (2), and Social Justice (1).

The Department of Social Work at the University of the Western Cape (UWC) offers three bachelor's degrees: a general programme, an extended version of the general programme, and a *Bachelor of Community Development*, in addition to master's and PhD programmes in Social Work (UWC 2024). Community Development is recognised as an independent occupational profession, rather than merely a specialisation within Social Work. Its mission is to empower local communities, strengthen individual and collective agency, and promote a participatory approach through stakeholder dialogue and the development of networks—ultimately contributing to greater community resilience. Supplementary social science courses provide critical insights into anthropology, social inequalities, crime, and violence (UWC 2024:89–91).

Social Service Workforce

As of 2024, nearly 40,000 social workers, 14,000 social auxiliary workers, and an additional 16,000 students in training were registered with the SACSSP (2024). The majority of practitioners are employed by non-governmental organisations (NGOs) and in the private sector, while a significant number remains unemployed (Gray and Lombard 2022:5; Chiwara 2024). Pressure on social workers, combined with rising costs, has led to the transfer of tasks to other agencies (Gray and Lombard 2022:2, 5). The *White Paper on Social Welfare* proposed the integration of social workers, child and youth care workers, and community development practitioners into a unified social service profession (Gray and Lombard 2022:2; Chiwara 2024). In addition to state-administered social assistance systems, communities and families often establish informal mechanisms to manage social risks (Mpedi 2017:47).

The founding of the National Association of Social Workers: South Africa (NASW: SA) in 2007 was intended to foster unity, improve the profession's image, and promote participation in the international dialogue through its membership in the IFSW. However, unity remained a challenge, and disputes ultimately lead to its dissolution (Gray and Lombard 2022:5).

3.4 Discussion

The country profiles indicate that social and community work have a significant role to play in reducing inequalities and addressing social exclusion. A rough calculation shows that 66 of the 248 SDG indicators are directly related to social services and social work. However, most of these indicators are not included in the 2024 Sustainable Development Report due to the lack of reliable data (Sachs et al. 2024). The Sustainable Development Report uses only two indicators to assess progress on reducing inequalities: the Gini coefficient and the Palma ratio. A third indicator—elderly poverty—was reported only for Germany (Sachs et al. 2024). While the Sustainable Development Report does not provide country-specific information, it does include a broader set of indicators that extend beyond purely economic measures (UN 2024). On the World Bank's SDG dashboard, 20 of the 36 key poverty indicators show 'no data available' (World Bank 2024). Lack of data makes it more difficult to tailor programmes and services to the needs of people and to hold governments accountable.

Germany, India, and Malaysia are characterised as tightly regulated societies (Gelfand 2018:24). In such contexts, social work has traditionally played a clearly defined role in controlling deviance—such as within the fields of criminal justice, child protection, and mental health—but has been less prominent in efforts to reduce social inequalities. Although both India and South Africa enshrine socio-economic rights within their constitutions, access to health services remains limited, and labour rights and access to justice are not consistently guaranteed (Sachs et al. 2024). For most countries, achieving greater equality remains an ongoing challenge and a work in progress.

A professional licencing system for social workers appears to be emerging globally, requiring institutionalised training, a complex set of skills, specialised knowledge, and the ability to engage in self-reflection (Oberlies 2022). Across all four countries studied, the self-perception of social work centres on supporting individuals and communities in their everyday lives. Social work is therefore regarded as the profession that 'goes alongside the people'.

Furthermore, all four countries are committed to expanding the number of trained social workers and improving the quality of their degree programmes. However, higher education can also increase the distance between social workers and marginalized communities and individuals in need. In some countries, such as India and Malaysia, degrees often serve as pathways to managerial careers rather than to frontline social work (UNICEF 2018:16). Inviting first-generation students, recruiting a diverse faculty, establishing rural campuses, and offering distance learning opportunities may be effective ways to keep professionals grounded.

All countries operate from a competence-oriented perspective; however, few prioritise outcomes, impact, and a process aimed at reducing social inequalities, addressing the actual needs of disadvantaged individuals and groups, and ensuring the participation of serviced communities (Association of Southeast Asian Nations [ASEAN] 2023).

While some subjects in social work education are universal—such as social work methods, humanities and social sciences, and basic research methods—the centrality of field exposure and meaningful practice continues to vary.

Often, fieldwork provides the only opportunity for students from middle-class backgrounds to encounter poverty, marginalisation, oppression, and 'other' cultures. The emphasis placed on research within social work programmes also differs; however, only original research can produce the body of knowledge that the profession requires. Furthermore, international exposure and transnational perspectives remain underdeveloped, even at the postgraduate level. Although exchange opportunities exist between the partner universities, Frankfurt remains distinctive for its focused transnational programme (Chmela 2024).

In the field of education, a striking injustice becomes apparent: students in higher income countries, such as Germany, benefit from free higher education and access to publicly funded jobs in social work, while students in middle and lower income countries must pay tuition fees and often struggle to secure employment. In such contexts, programmes may be shaped more by the demands of paying students than by the need to provide high-quality services.

While South Asian countries prioritise governmental services, the German principle of subsidiarity gives precedence to NGOs in service delivery. There is evidence that NGOs make a meaningful difference, not least because they invest an average of 500,000 USD in their respective fields of work (Shakoor et al. 2024). Nevertheless, the position of NGOs remains fragile.

- Voluntary work and social movements have a long tradition in India (Working Group of Professional Social Workers 2021), tracing back to tribal movements and the establishment of non-governmental agencies after independence (Shareef 2023). NGOs have since become major drivers of public participation and poverty alleviation. Nearly 200,000 NGOs are currently registered in the government database. Funding comes from various sources, including development aid, domestic charity foundations, and, more recently, Corporate Social Responsibility (CSR) initiatives.
- NGOs were first established in Malaysia around 1970, focusing on consumer, women, and environmental rights. Rights-based NGOs have been particularly prominent and were historically classified as either 'political' or 'friendly' (Khoo 2013). Currently 95,694 non-political NGOs are registered across 11 categories, including welfare and charity, social, human rights, and mutual benefit, to mention a few (New Straits Times 2024). While NGOs are now considered influential actors, securing adequate funding remains a persistent challenge (Ahmad and Arshad 2022).
- In South Africa, it has been noted that 'no discussion of poverty, equality, or development today will be complete without considering the role of NGOs' (Makofane and Selepe 2022:262). Approximately 270,000 NGOs are registered with the Nonprofit Organisation (NPO) Directorate in the Department of Social Development, of which 25,700 are classified as public benefit organisations (PBOs), employing nearly one million people (Department of Social Development 2023). Nevertheless, funding remains a persistent challenge (Makofane and Selepe 2022:264).
- In Germany, 'free sponsors' (NGOs) date back to 1923, when the youth welfare system was first codified into law and churches still played a central role in education. Today, the principle of subsidiarity ensures a plurality of services and allows clients to choose among them (Sec. 5 SGB 8). Approximately 29,600 are officially recognised as providers of youth welfare services, employing nearly 200,000 employees—2.5 times more than in public institutions (DeSTATIS 2024c). Once a service is included in the local planning, it can be contracted and financed through public funds (Sec. 77 SGB 8).

This overview shows that NGOs and their staff play a dominant role in providing social services in all four countries. Their workforce typically consists of both governmental and non-governmental services for children, youth, adults, the elderly, families, and communities (GSSWA 2019:13). *Professionals* hold a degree and, where required, a licence in Social Work; *paraprofessionals* support service delivery where systems are underdeveloped or overstretched; and *allied workers* are engaged in related sectors such as education, health, and justice (GSSWA 2019:16).

The composition of the workforce reflects the specific gaps and needs of each country. In Malaysia, professional social workers are primarily trained for government services. India and South Africa have a highly layered workforce, with a strong focus on community work. Germany, by contrast, has regulated sectors—such as youth services, probation offices, and legal assistance—where only trained and licenced social workers are permitted to work.

All countries appear to be implementing initiatives aimed at increasing the number of well-trained professionals. During the COVID-19 pandemic, it became evident that social care is among the 'essential jobs', especially in light of weakening family structures and shifting demographics. Moreover, the social services workforce creates important employment opportunities for women and migrants (Nivorozhkin and Poeschel 2022).

Professional associations at both national and supra-national levels play a significant role in shaping professional identity, debating ethical principles, implementing codes of service, and engaging with critical issues such as the reduction of social inequalities, poverty alleviation, and sustainability.

3.5 Conclusion

The desk research revealed a lack of data on inequalities in official reports. Data collection tends to focus on economic indicators, is often outdated, and primarily relies on government sources. The call for improved data collection (SDG 17.18) is therefore well-founded, though not all countries may be able to meet what is arguably an overly ambitious target.

Social work and civil society can play a vital role in contributing necessary data towards achieving the goals:

- Applied research in social work should focus on assessing the impact of social interventions in reducing inequalities.
- Social research—particularly on poverty, inclusion, discrimination, and migration—should be made openly accessible and ideally indexed in a centralised directory.
- Data collected through baseline studies conducted by local NGOs for international donors should be more standardised and compiled to contribute to a broader, more comprehensive picture.

Small measures can enhance collaboration among the stakeholders in the field, focus on outcomes for the people, and significantly increase the impact of the findings. The involvement of social research and civil society will make a difference.

In addition, the SDG 10 targets provide a helpful framework for evaluating laws, policies, and programmes regarding social services (or the lack thereof).

- Does the measure reduce poverty, and prevent the risk of becoming poor(er)?
- Does it include all? Does it give equal opportunities to all?
- Is the law, policy, or programme—directly or indirectly—discriminating against groups or individuals? Does it produce equal outcomes?
- Do the combined measures lead to ever greater equality?
- Does any related measure make migration and migrants safer?

When it comes to legislation involving social services, the role of social work in delivering for the people must be clearly discussed and defined. Students of social work must be equipped to understand social 'problems', forms of oppression, and their effects on individuals, communities, and society. An SDG-sensitive curriculum, combined with praxis and research, can help lead the way forward:

- Help reduce poverty and hunger (SDGs 1 and 2), and prevent unemployment and exploitation (SDG 8).
- Provide access to health and counselling services (SDG 3) and protect children, women, and other vulnerable groups from violence (SDGs 5 and 16.2).
- Promote greater equality, inclusion, and non-discrimination; support integration efforts (SDG 10); ensure equal opportunities and education for all (SDG 4); and advance gender equality (SDG 5).
- Contribute to mitigating environmental and climate challenges (SDGs 6, 7, 13, 14, and 15).
- Engage in international solidarity (SDGs 16 and 17).

While many topics are already part of social work curricula, it is important to further incorporate and develop environmental justice. Indigenous concepts can support reflection on climate change and its environmental impacts. Environmental and climate concerns should inform the content and practice of social work, as well as its professional ethics and policy advocacy.

Syllabi and reading lists—both in the Global South and the Global North—must include literature from both regions (Singh and Saumya 2021:6). However, much of the 'universal' material still originates from Anglo-American sources (Pandey and Kumar 2018:202). Some students may struggle with English-language materials, but 'broken English' remains the only truly global language. Importantly, indigenous perspectives offer valuable lessons for the Global North to regard the Earth as a vital community member rather than merely a resource.

In future, social work must embed the idea of global interconnectedness and sustainability beyond the boundaries of the nation-state. International exchange is crucial for experiencing culture shocks, relating to the feelings of otherness, and critically examining Western beliefs and privileges (Ali 2022). Such experiences can ultimately benefit social workers, their clients, and broader society. They foster the solidarity, partnerships, and cooperation necessary for addressing global challenges.

While the importance of the social service workforce is obvious, the role and function of social workers are often dubious. Social work should insist on being a 'timid' profession (Schütze) that deals with complexities and paradoxes when supporting individuals, groups, and communities in their everyday struggles. To do so, the composition of the workforce must include trained locals (paraprofessionals), licenced social workers, and professionals with expert knowledge, while also encouraging collaboration within multiprofessional teams. In this context, the vital role of NGOs in maintaining close connections with communities cannot be overstated.

Achieving these goals also requires the active involvement of 'pressure groups' that advocate for the SDGs in general, and for strengthened social services to reduce inequalities in particular. This calls for the establishment of networks and the creation of structures—such as associations—at local, national, and global levels.

References

Actovin CA, Prasad S (2023) Indic values and ethics of social work profession of India. J Soc Work Soc Dev 14(2):123–131

Adaikalam F (2014) Contextualising social work education in India. Altern Cuad Trab Soc 21(21):216–232

Afrobarometer (2023) South Africans see gender-based violence as most important women's-rights issue to address. https://www.afrobarometer.org/wp-content/uploads/2023/11/AD738-South-Africans-see-gender-based-violence-as-a-top-priority-Afrobarometer-24nov23.pdf. Accessed 14 July 2023

Ahmad Z, Arshad R (2022) NGOs and community: a case study of roles of NGOs in Malaysia. Adv Econ Bus Manag Res 209:319–322

Ali W (2022) The shifting landscape of international practicum in social work education. Int J Soc Work 9(1):1

Amin SM, Ahmad Suhaimi SS, Nazuri NS (2022) Social work in Malaysia: the history and development. Int J Acad Res Bus Soc Sci 12(12):2359–2369

Amin SM, Mohamad MS, Ahmad TLT (2019) The development of social work undergraduate programs in Malaysia public universities: issues, challenges and opportunities. Pertanika J Soc Sci Humanit 27:45–59

Antidiskriminierungsstelle des Bundes (2022) Annual report 2022. Federal Anti-Discrimination Agency, Berlin

Association of Southeast Asian Nations (ASEAN) (2023) Regional guidance on the development and use of social service workforce to population ratios in ASEAN. https://asean.org/wp-content/uploads/2024/06/Development-and-Use-of-Social-Service-Workforce-to-Population-Ratios-in-ASEAN_FINAL.pdf. Accessed 30 July 2024

Azizul MDA, Amin SM (2019) Direction of social work education in Malaysia: strategies to implement good governance. IOSR J Humanit Soc Sci 24(8):42–51

Bundesamt für Arbeit (2024) 2.7 Sozialwesen. https://statistik.arbeitsagentur.de/DE/Statischer-Content/Statistiken/Themen-im-Fokus/Berufe/AkademikerInnen/Berufsgruppen/Generische-Publikationen/2-7-Sozialwesen.pdf;jsessionid=81DDFCD451E9C7A481C62B3A29AC1021?__blob=publicationFile&v=9. Accessed 15 July 2024

Bundesministerium für Arbeit und Soziales (2021) Report on poverty and wealth. https://www.armuts-und-reichtumsbericht.de/SharedDocs/Downloads/Berichte/sixth-report-on-poverty-and-wealth.pdf?__blob=publicationFile&v=2. Accessed 30 July 2024

Bundestag D (2014) Rechtsprechung zum Sozialstaatsprinzip und zu sozialen Grundrechten in Deutschland. https://www.bundestag.de/resource/blob/406948/da4397e0f0e483547f7e1dcccd59f38c/WD-6-200-14-pdf-data.pdf. Accessed 14 July 2024

Chiwara P (2024) Structural inequalities in Namibia and South Africa: a critical social work perspective. South Afr J Soc Work Soc Dev 36(2):1–18

Chmela A (2024) Social work education towards the 2030 agenda for sustainable development: a student account from Germany. Soc Dialogue Mag 28:38–41

Council on Higher Education (CHE) (2015) Higher Education Qualifications Sub-Framework: Qualification Standards for Bachelor of Social Work. http://www.che.ac.za/publications/standard-reviews/qualification-standards-bachelor-social-work. Accessed 14 July 2024

Das TK, Mehta B, Nikku BR et al (2022) Social work education decolonizing south Asian social work: an analysis. Soc Work Educ 41(5):601–619

Dash BM (2017) Revisiting eight decades of social work education in India. Asian Soc Work Policy Rev 11(1):66–75

Department of Social Development (2023) Minister Lindiwe Zulu applauds the critical role played by non-governmental organisations. [Media statement] 27 February. Available at: https://www.dsd.gov.za/index.php/21-latest-news/463-minister-lindiwe-zulu-applauds-the-critical-role-played-by-non-governmental-organisations. Accessed 4 May 2025

Desai M, Singh R, Bhatt S (eds) (2020) Journeys in social work education in India: emerging reflections. Rawat Publications, Jaipur

DeSTATIS (2021) Discrimination at work. https://www.destatis.de/EN/Themes/Labour/Labour-Market/Quality-Employment/Dimension7/7_2_DiscriminationWork.html. Accessed 30 July 2024

DeSTATIS (2024a) Tabelle. Studierende: Deutschland, Semester, Nationalität. https://www-genesis.destatis.de/genesis/online?sequenz=tabelleErgebnis&selectionname=21311-0003#abreadcrumb. Accessed 15 July 2024

DeSTATIS (2024b) Pressemitteilung Nr. N004 vom 24. Januar 2024. https://www.destatis.de/DE/Presse/Pressemitteilungen/2024/01/PD24_N004_p002.html. Accessed 15 July 2024

DeSTATIS (2024c) Presse. 29 600 Träger der Kinder- und Jugendhilfe im Jahr 2022. https://www.destatis.de/DE/Presse/Pressemitteilungen/2024/03/PD24_129_225.html. Accessed Apr 2024

Deutsche Presse-Agentur (DPA) (2023) Angeblicher Personalschlüssel in der Flüchtlingsbetreuung weit übertrieben. https://dpa-factchecking.com/germany/230123-99-327124/. Accessed 15 July 2024

Deutscher Bundestag (2012) Bericht zur Situation der Frauenhäuser, der Fachberatungsstellen und anderer Unterstützungsangebote für gewaltbetroffene Frauen und deren Kinder. https://dserver.bundestag.de/btd/17/105/1710500.pdf. Accessed 15 July 2024

Economic Planning Unit (2021) Twelfth Malaysian Plan 2021–2025: A prosperous, inclusive, sustainable Malaysia. https://pulse.icdm.com.my/wp-content/uploads/2021/09/Twelfth-Plan-Document_compressed-1.pdf. Accessed 14 July 2024

Fuziah S, Teoh AH (2017) Professionalization of social work in Malaysia through legislation: a literature discussion on concepts, issues and challenges. Jurnal Pembangunan Sosial 20:117–134

Gelfand M (2018) Rule makers, rule breakers: how tight and loose cultures wire our world. Simon & Schuster, New York

Ghatak M, Rishabh K (2024) Poverty in India over the last decade: Data, debates, and doubts. https://www.theindiaforum.in/economy/poverty-india-over-last-decade. Accessed 14 Apr 2024

Global Centre for Pluralism (2023) Global pluralism monitor: South Africa. https://monitor.pluralism.ca/wp-content/uploads/2023/06/GPM-Report-South-Africa-EN-WEB.pdf. Accessed 14 July 2024

Global Social Service Workforce Alliance (GSSWA) (2019) Social service workforce mapping toolkit. GSSWA, Washington, DC

Gray M, Lombard A (2022) Progress of the social service professions in South Africa's developmental social welfare system: social work, and child and youth care work. Int J Soc Work 32(4):429–441

Harini V, Suresh B, Kumar R (2023) Influence of social work education on recognition of social work as a profession in India. Juni Khyat UGC Care Group I Listed J 13(12):176–181

Hatta ZA, Ali I, Subramaniam J et al (2014) Professional and functional alternative social workers: a case study of Malaysia. Asian Soc Work Policy Rev 8(2):135–155

Hochschulkompass (2025) Advanced Study Programme Search: Social Work (Master's) at German Universities. Available at: https://www.hochschulkompass.de/studium/studiengangsuche/erweiterte-studiengangsuche.html?tx_szhrksearch_pi1%5Bsearch%5D=1&tx_szhrksearch_pi1%5Bstudtyp%5D=3&tx_szhrksearch_pi1%5BQUICK%5D=1&tx_szhrksearch_pi1%5Bfach%5D=Soziale+Arbeit+. Accessed 4 May 2025

Hua TA (2023) Professionalisation of social work in Malaysia: 1969–2010. Universiti Utara Malaysia, Sintok

Hua TA, Shaffie F (2016) Social work professionalism: conflicting discourse or opportunity to the profession and to the government? The case of Malaysia and its implication to the ASEAN region. ASEAN Soc Work J 4(1):65–76

International Federation of Social Workers (IFSW) (2017) Social work and sustainable development: World Social Work Day at the UN in Geneva, 2017 and 2018. https://www.ifsw.org/wp-content/uploads/ifsw-cdn/assets/ifsw_83551-10.pdf. Accessed 10 July 2024

International Organisation for Migration (IOM) (2024) Malaysia. https://www.iom.int/countries/malaysia. Accessed 14 July 2024

Jayasooria D (2016) Sustainable development goals and social work: opportunities and challenges for social work practice in Malaysia. J Hum Rights Soc Work 1(1):19–29

Jabatan Kebajikan Masyarakat Malaysia [JKM] (2020) Pelan strategik Jabatan Kebajikan Masyarakat 2021–2025

Kaltenborn M, Krajewski M, Kuhn H (eds) (2020) Sustainable development goals and human rights. Springer Nature, Cham

Khoo YH (2013) The NGO-government relations in Malaysia: historical context and contemporary discourse. Malays J Democr Election Stud:76–85

Khosla M (2010) Making social rights conditional: lessons from India. Int J Const Law 8(4):739–765

Koutronas E (2020) Social protection for all: the next step for the Malaysian welfare. West Virginia University, West Virginia

Landgericht Frankfurt am Main (2022) Jahresbericht 2021 der Sozialen Dienste der Justiz. https://ordentliche-gerichtsbarkeit.hessen.de/sites/ordentliche-gerichtsbarkeit.hessen.de/files/2022-05/bfbg_jahresbericht_der_sozialen_dienste_2021.pdf. Accessed 15 July 2024

Leburu-Masigo G (2020) Women and South Africa's triple challenge. Soc Work Maatskaplike Werk 56(2):175–185

Ling H-K (2002) Social work practice in Malaysia: a vision of direction. In: Tan NT, Dodds I (eds) Social work around the world II. International Federation of Social Workers (IFSW) Press, Berne, pp 83–97

Lombard A, Grobbelaar M, Pruis S (2003) Standards for social work qualifications in South Africa. Soc Work Maatskaplike Werk 39(1)

Lombard A, Twikirize JM (2014) Promoting social and economic equality: social workers' contribution to social justice and social development in South Africa and Uganda. Int Soc Work 57(4):313–325

Makofane HR, Selepe M (2022) The role of non-governmental organizations (NGOs) in South Africa with specific reference to rural development. Eurasian J Soc Sci 10(4):262–276

Malaysian Association of Social Workers (MASW) (2024) To seek professional competency in social work practice, education and research. https://www.masw.org.my. Accessed 14 June 2024

Mathew A (2020) India: Modi government circulates draft Bill to regulate autonomy in social work practice, education. https://www.nationalheraldindia.com/india/modi-government-circulates-draft-bill-to-regulate-autonomy-in-social-work-practice-education. Accessed July 2024

McAuliffe M, Oucho L (2024) World migration report 2024. International Organization for Migration, Geneva

Ministry of Health Malaysia (2020) Strategic framework of the medical programme: Ministry of Health Malaysia 2021–2025. Hospital Management Unit, Medical Development Division, Kuala Lumpur

Mohammed AA, Uddin MS, Bougangue B (2020) Starting the indigenisation process of social work practice in Malaysia. Soc Work Soc 18(1):1–14

Mpedi LG (2017) Social protection law in the Republic of South Africa. Recht Afr 20(1):33–75

National Council of Social Work (NCSW) (2020) Draft of National Council of Social Work (education and practice) bill. https://ourgovdotin.wordpress.com/wp-content/uploads/2020/11/draft-of-national-council-of-social-work-education-and-practice-bill-2020-5-10-20.pdf. Accessed 15 July 2024

National Curriculum Resource Group (2022) Report of the National Curricular Framework for Social Work Education. https://www.napswi.org/pdf/Report_National_curricular_framework_by_NCC_NCSWE.pdf. Accessed 15 July 2024

New Straits Times (2024) 95,694 NGOs registered in Malaysia, excluding political NGOs. 3 May. Available at: https://www.nst.com.my/news/nation/2024/05/1055959/95694-ngos-registered-malaysia-excluding-political-ngos. Accessed 4 May 2025

Nikku BR, Pulla V (2014) Global agenda for social work and social development: voices of the social work educators from Asia. Int Soc Work 57(4):373–385

Nivorozhkin A, Poeschel F (2022) Working conditions in essential occupations and the role of migrants. Econ Anal Policy 74:250–261

Noyoo N, Seepamore B, Ncube M et al (2021) Evolution of social welfare and social work in South Africa. In: Noyoo N (ed) Social welfare and social work in southern Africa. African Sun Media, Stellenbosch, pp 359–369

Oberlies D (2022) Rechtlich fundierte Sozialarbeit—ein Beitrag zum Professionsverständnis. https://fhffm.bsz-bw.de/frontdoor/deliver/index/docId/6505/file/Zum_Abschied_Sozialarbeit_final.pdf. Accessed 10 July 2024

Office of the High Commissioner for Human Rights (OHCHR) (2024) UN treaty body database. https://tbinternet.ohchr.org/_layouts/15/TreatyBodyExternal/Treaty.aspx?Lang=en. Accessed 14 July 2024

Oxfam India (2022) The India discrimination report 2022. Oxfam India, New Delhi

Pandey UC, Kumar C (2018) A SDG compliant curriculum framework for social work education: issues and challenges. In: Filho WL (ed) Implementing sustainability in the curriculum of universities: approaches, methods and projects. Springer International Publishing, Cham, pp 193–206

Parker J, Crabtree SA, Azman A et al (2017) Towards a typology of social work assessments: developing practice in Malaysia, Nepal. United Kingdom and Vietnam Soc Work Soc 15(2)

Prasad S, Actovin CA, Dash BM et al (2024) Changing trends of social work curriculum: a study of selected social work institutions in India. Afr J Soc Work 14(2):100–107

Pulver C, Matti T (2021) Soziodemografische Herkunft, Persönlichkeitsmerkmale und Studienwahlmotive von Studierenden der Sozialen Arbeit: Anregungen und Hinweise für die Ausgestaltung der Förderung der Persönlichkeitsentwicklung in der Hochschulausbildung. Z Bildungsforsch 11(3):403–420

Rai V, Mishra RC (2024) Muslims and social work institutions in India. Int Res J Adv Eng Manag 2(6):1840–1844

Ravallion M (2020) Ethnic inequality and poverty in Malaysia since May 1969. Part 1: inequality. J Dev Econ 134:105040

Rehadat S (2022) Bundesarbeitsgemeinschaft Werkstätten für behinderte Menschen. https://www.rehadat-statistik.de/statistiken/berufliche-teilhabe/wfbm/bagwfbm/. Accessed 15 July 2024

Sachs JD, Lafortune G, Fuller G (2024) The SDGs and the UN summit of the future: sustainable development report 2024. Dublin University Press, Dublin

Sahu KK (2017) Mental health act 2017 and psychiatric social work practice. Indian J Psychiatr Soc Work 8(2):1–2

Sakala BM, Mabundza LP, Noyoo N (2022) An exploration of social assistance programmes in Eswatini, South Africa and Zambia. South Afr J Soc Work Soc Dev 34(1):1–14

Schäfer P, Bartosch U (2016) Qualifikationsrahmen Soziale Arbeit. https://www.fbts-ev.de/qualifikationsrahmen-soziale-arbeit. Accessed 15 July 2024

Seth N (2018) Social services and the SDGs: 9th Conference for the Social Work and Sustainable Development Goals, Sharjah, UAE, Monday 23 April 2018. https://unitar.org/sites/default/files/media/file/social_services_and_the_sdgs-April2018.pdf. Accessed 3 June 2024

Shakoor DA, Farooq CIT, Khan AK et al (2024) Role of NGOs in shaping social development. Int J Contemp Issues Soc Sci 3(1):831–838

Shareef DMS (2023) Role and functions of NGOs in social development—an overview. Int J Multidiscip Educ Res 12(8):85–90

Singh T (2021) Remodelling the curriculum of social work in India. Soc Work Educ 42(7):1057–1073

Singh T, Saumya S (2021) Social work education in India: old and new. Soc Work Educ 41(6):1089–1108

Solano G, Huddleston T (2020a) Measuring policies to integrate migrants across six continents. https://www.mipex.eu. Accessed June 2024

Solano G, Huddleston T (2020b) Germany: Key findings. https://www.mipex.eu/germany. Accessed May 2024

Solano G, Huddleston T (2020c) India: Key findings. https://www.mipex.eu/india. Accessed 15 May 2024

Solano G, Huddleston T (2020d) South Africa: Key findings. https://www.mipex.eu/south-africa. Accessed 14 July 2024

South African Council for Social Service Professions (SACSSP) (2024) Social work. https://www.sacssp.co.za/for-the-profession-social-work/. Accessed 14 July 2024

South African Qualifications Authority (SAQA) (n.d.) Registered qualification that has passed the end date: Bachelor of Social Work. https://regqs.saqa.org.za/viewQualification.php?id=23994. Accessed 14 July 2024

Statista (2024) Berufe mit den größten Mangel an Fachkräften in Deutschland im Jahr 2022. https://de.statista.com/statistik/daten/studie/1326564/umfrage/berufe-mit-den-groessten-fachkraefteluecken/. Accessed 15 July 2024

Tata Institute of Social Science (TISS) (2024a School of Social Work—Mumbai Campus: Centres. https://tiss.edu/centres/mumbai-campus/school-of-social-work/. Accessed 20 July 2024

Tata Institute of Social Science (TISS) (2024b) PG programmes. https://admissions.tiss.edu/admissions/ma/programmes/. Accessed 21 July 2024

Tata Institute of Social Science (TISS) (2024c) Under graduate programmes: Bachelor of Arts in Social Work. https://admissions.tiss.edu/view/6/admissions/ba-ma-admissions/about-ba-programmes/#BSW_M. Accessed 15 July 2024

Tata Institute of Social Science (TISS) (2023) Yearly status report—2021-2022. https://tiss.ac.in/uploads/2021-22.pdf. Accessed 15 July 2024

UN Women (2018) SDG indicator dashboard. https://data.unwomen.org/data-portal/sdg. Accessed 14 July 2024

United Nations (UN) (2024) The sustainable development goals report 2024. United Nations, New York

United Nations Children's Fund (UNICEF) (2018) State of social service workforce in South Asia. UNICEF ROSA, Kathmandu

United Nations Children's Fund (UNICEF), Malaysian Association of Social Workers (MASW) (2023) Fact sheet: Professionalizing social work in Malaysia: The case for the Social Work Profession Bill. https://www.unicef.org/malaysia/media/4096/file/HAU%20Social%20Workers%20Factsheet_2023.pdf. Accessed 12 May 2024

United Nations Educational, Scientific and Cultural Organisation (UNESCO) (2016) World social science report 2016: challenging inequalities: pathways to a just world. International Social Science Council; University of Sussex, Paris

University of the Western Cape (UWC) (2024) Department of Social Work. https://www.uwc.ac.za/study/all-areas-of-study/departments/department-of-social-work/overview. Accessed 14 July 2024

Vareed BP (2023) Dimensions of critical social work practice in India. J Soc Work Educ Pract 8(3):1–11

Working Group of Professional Social Workers (2021) Campaign for National Council for Social Work Education. https://napswi.org/pdf/National_Council_on_SW_Education-2021.pdf. Accessed 15 July 2024

World Bank (2024) World. http://worldbank.sdgsdashboard.org/home. Accessed 24 July 2024

World Bank (2024a) Refugee population by country or territory of asylum. https://data.worldbank.org/indicator/SM.POP.REFG?view=map. Accessed 14 July 2024

World Bank (2024b) Poverty and equity briefs. https://www.worldbank.org/en/topic/poverty/publication/poverty-and-equity-briefs. Accessed 14 July 2024

Role of Social Work Practice: The Local to the Global for Advancing the Sustainable Development Goals

4

Shewli Kumar and Meera Suresh Babu

Abstract

The central promise of the 2030 Agenda for Sustainable Development is to 'leave no one behind'. This commitment is closely aligned with the core values of the social work profession, including human rights, social justice, and equality. In the face of rapidly intensifying global challenges, the Sustainable Development Goals (SDG) offer a vital framework for social workers to address deep-rooted inequalities. Understanding how social workers perceive and experience social inequalities and discriminatory practices—both within and among countries—is essential to advancing this agenda. Adopting a qualitative research design, the study involved 16 in-depth interviews and four focus group discussions with social work practitioners from Germany, Malaysia, South Africa, and India. The findings reveal that the diversity of practices and methods employed by social work practitioners is significantly influenced by the nature of the welfare state in their respective countries. Social workers operating within state systems often feel constrained by bureaucratic regulations and administrative procedures, which hinder their ability to address the structural roots of poverty, discrimination, and social inequality. However, practitioners working with non-governmental organisations and movements are often better positioned to engage directly with communities and address discriminatory practices effectively. The transnational approach for addressing social inequalities has a greater potential to address the SDGS, especially Goal 10.

Keywords

Sustainable development goals · Inequalities and discriminatory practices · Social work · Germany · Malaysia · South Africa · India

4.1 Introduction

Social work is a 'profession' grounded in core values that promote social justice and equality in an unjust and inequitable world. Historically, it emerged as a 'helping' profession, working in partnership with the state to provide services for the poor and marginalised. As countries experienced shifts in their developmental trajectories, these changes influenced their course content, pedagogy, and practice of social work. In response to evolving government welfare initiatives, different streams of thought emerged within social work, shifting emphasis from a generic focus to specialisation.

S. Kumar (✉)
School of Social Work, Tata Institute of Social Sciences, Mumbai, India
e-mail: shewli4@tiss.ac.in

M. S. Babu
School of Social Work, Tata Institute of Social Sciences, Mumbai, India

N. V. Roman et al. (eds.), *Social Inequalities Within and Among Countries*, Sustainable Development Goals Series, https://doi.org/10.1007/978-3-032-08741-6_4

Over time, social work practice and emerging concerns within the developmental states of the Global South prompted a critical questioning of Western and Eurocentric approaches to social work education and practice. This gave rise to a new strand of social work theorising and field-based interventions focused on the decolonisation of social work education (Gray et al. 2013). There is a growing push to develop social work theory and practice from indigenous perspectives, grounded in writings and research that seek to examine and intervene in the social world through the experiences of diverse individuals relegated to the margins. This movement emphasises the reclamation of indigenous knowledge systems and their integration into both research and practice. The following definition of 'social work' provided by the International Association of Schools of Social Work (IASSW 2024) resonates with these ideas:

> Social work as a practice-based profession and an academic discipline aims to promote social change and development, social cohesion, and the empowerment and liberation of people. It is guided by the principles of social justice, human rights, collective responsibility and respect for diversities and informed by theories of social work, social sciences, humanities and indigenous knowledge. Social work engages people and structures to address life challenges and enhance wellbeing. The above definition may be amplified at national and/or regional levels.[1]

Social work engages with people and structures to address life challenges and enhance well-being. Its principles are closely aligned with those of the Sustainable Development Goals (SDGs). SDG 10, which aims to reduce inequality within and among countries, is of prime importance to social work practitioners worldwide. Social workers adapt various methods according to the context, integrating theoretical insights into their practice.

The latest Global Agenda for Social Work (IASSW 2024), as defined by the International Federation of Social Workers (IFSW), emphasises the need to address a world in crisis, with thousands of people dying due to violent conflict, millions displaced by wars, climate change, environmental destruction, and persisting poverty and inequality. Governments are promoting 'competition over collaboration and sovereignty over solidarity'. Social work needs to adopt new strategies, approaches, actions, actors, movements, and platforms for transformative change, spanning from micro to macro practice, to bring about structural change.[2] This requires co-creating and co-building actions, policies, practices, and support for the eco-social world, affirming inclusion, intergenerational respect, peace, upholding human rights, and sustainable development.

This chapter presents the findings of a qualitative research study on the experiences of social work practitioners in their chosen sectoral areas across four countries: Germany, Malaysia, South Africa, and India. It also explores their understanding of the SDGs, specifically SDG 10, concerning inequalities and discrimination within and among countries. The chapter is divided into three parts. The first part includes a literature review of studies conducted on social work practice in the four countries, primarily contextualising them within India, Germany, Malaysia, and South Africa while bringing in a transnational perspective for the research. The second section explains the research methodology. The third section presents the findings, which highlight social work practitioners' knowledge regarding the core elements of SDG 10 and the various ways through which SDGs are or can be localised through social work practices and interventions.

4.2 Social Work Practice in the Four Countries: A Literature Review

This section organises selected literature on social work practice to trace its historical trajectory and examine the current state of practice in the selected countries.

[1] IASSW 2024, https://www.iassw-aiets.org/global-definition-of-social-work-review-of-the-global-definition/ Accessed 19 Apr 2025.

[2] https://www.icsw.org/images/docs/News/2023/11/GlobalAgenda2030-Theme2024.pdf

4.2.1 Historical Development of Professional Social Work in Germany, Malaysia, India, and South Africa

The history of professional social work education and practice has its roots in the emergent socio-economic problems and deprivations brought about by rapid industrialisation in nineteenth-century Western countries. Industrialisation led to mass poverty, hunger, and the neglect of specific groups of people. Simultaneously, there was a shift in world systems, marked by the fall of monarchies and the emergence of democratic welfare states. In Germany, poverty and misery were questioned as unnatural and necessitated government intervention. The significant problems unleashed by increasing capitalism and industrialisation were conceptualised as *Soziale Frage* (social issue) (Hammerschmidt et al. 2017). Shifts in national governance systems, the World Wars, and the collectivisation of labour, women, and youth significantly contributed to the development of social work as a field of practice in Germany. A key figure in this evolution was Alice Salomon, a pioneer of the women's movement, who founded the soziale Frauenschule (the Social School for Women). This institution marked the first formal educational initiative in social work and stands as a major milestone in the history of professional social work in Germany (Wendt 2017). Throughout its history, social work education and the profession underwent several historical moments and tensions, including complex interactions with Nazi Germany. Following World War II, social work was reclaimed and refashioned as the country transitioned towards a conservative and corporatist welfare regime. The need to professionalise social work was recognised, and in 1989, the German Association of Social Work was founded, along with several other associations (Laging et al. 2021). There was ongoing tension between academic social pedagogy—the theory of science—and social work practice. This was finally resolved around the year 2000, when all groups agreed to adopt the generic term *Soziale Arbeit* for both *Sozialearbeit* (social work) and *Sozialepadagogik* (social pedagogy).

Social work education and practice were introduced in Malaysia after World War II, beginning in 1946 with the establishment of the Department of Social Welfare (Mohammad and Uddin 2020). Rooted in British colonial legacy, the field was initially developed by medical professionals in 1955. In the 1970s, medical social workers established the Malaysian Association of Social Workers (MASW), which was officially registered as an organisation in 1973. The trend of training professional social workers in Western theories and methods continues to influence education and practice in Malaysia. This has been analysed and critiqued by social workers themselves over time. Currently, the profession applies a biopsychosocial approach to interventions aimed at solving clients' problems. However, there exists a second strand of social workers who focus on social change, social policy, and legislative reforms (Hatta et al. 2014). In Malaysia, as in India, voluntary social workers continue to operate alongside professional social workers. Known as functional alternative social workers, these volunteers lack formal training but have a long history of serving the underprivileged across various sectors. Both professional and functional social workers have served as social welfare officers, child protection officers, probation officers, social work lecturers, and medical social workers for many years. This coexistence of trained and untrained social workers has created a contested space, and the MASW is working diligently towards establishing a professional licencing system by advocating for the formation of a Malaysian Social Work Council.

India has a trajectory for the introduction of professional social work training and practice that is quite similar to that of Malaysia. Historically, like Germany and Malaysia, volunteer workers emerged from various religious communities to alleviate the suffering and misery of the poor and deprived people in the country. With the advent of the British colonial regime in India, discussions began around trained social work, and interventions were designed similarly to the settlement houses in Britain in the 1920s (Mishra 2021). By this time, the nationalist movement was gaining momentum in India, and several social service organisations began working in rural areas, engaging in legal activism and other initiatives. A close-knit group of social service workers (as they were known then) emerged, coming together for conferences and meetings.

In 1923, the Bombay Council of Social Workers was formed, which held several conferences where the need for trained social workers was frequently discussed. Numerous national leaders initiated social reforms that contributed to the call for freedom and independence from British rule. Ironically, however, the first school of social work was established in 1936, focusing on the Western model of social work as envisioned by Clifford Manshardt, leading to a more standardised and procedural approach to the field (Singh et al. 2011). Social work education began with the Tata Institute of Social Sciences offering professional social work training in 1936. However, social workers had to respond to the post-independence devastation, which was accompanied by the violent partition of the country and the proliferation of refugee camps housing destitute people, widows, women, and children. At that time, Gandhian ideas of social reconstruction informed social work practice, emphasising rural social work, community work, and family and child welfare. New methods of social work were conceptualised, such as social action and advocacy, which became essential for organising people from the margins and interfacing with the government for social change and justice (Gore 1997).

The traditional methods of casework and group work have since been critically analysed and questioned by social work educators and practitioners. Today, in India, a wide range of social work approaches and methods has emerged, characterised by consistent engagement with state systems (health care, criminal justice systems, child protection systems, and mental health systems), social movements, and community engagements. More significantly, numerous trained social workers in diverse organisations are working towards anti-oppressive and emancipatory approaches, addressing the intersectionality of gender, caste, tribe, minority, and other marginalised social groups. Decoloniality and the reclamation of indigenous approaches inform this kind of social work practice in India, which interfaces with social groups and communities created by those outside the social work profession (Bodhi 2020).

In South Africa, social work education was introduced in the 1930s, with the Jan Hofmeyer College being the first institution to offer formal training in the field. According to Chetty (1999), social relations in the country, since the arrival of Europeans in the nineteenth century, have been marked by patterns of domination and subordination, primarily shaped by class and race. Political and social relations continue to impact and perpetuate exploitation and exclusionary inequalities within society. These issues have persisted in social relations and pose a challenge for social workers to address. Initially, welfare work was initiated through racially segregated initiatives, which meant that the benefits largely went to white people, while discrimination based on race and identity continued to prevail. Post-apartheid efforts to integrate social work organisations included the establishment of the South African Council for Social Work in 1989. However, as the social work profession progressed in the country, debates regarding racial discrimination, discrimination against indigenous peoples, and the treatment of Indians brought in as indentured labour remained a reality. Social work educators and practitioners had to acknowledge that working within the state systems of welfare did not necessarily address social inequalities, poverty, and discrimination (Smith 2014).

The period from 1970 to 1980 was particularly interesting for social work practice in all four countries. During the 1970s, social movements, self-help groups, and women's movements influenced German social work thinking and interventions. Simultaneously, in South Africa (Willmore 1985; Gore and Gore 1977; Bodhi 2011), and Malaysia (Mohammad and Uddin 2020), critical questions emerged regarding the imposition of Western theories and methods of social work practice. These were largely considered remedial and clinical, with social work being limited to making people fit and adjust to their social conditions rather than changing them. Such approaches were not based on the contextual realities of people's lives in different countries, particularly those with experiences of colonial rule. The arguments called for a shift towards developmental work that embraced a heterogeneity of approaches and focused on creating solidarity, freedom, equality, and justice.

4.2.2 Neoliberalism, Shifts in Welfare States and Social Work Practice

With the shifts in the world economy towards neoliberalism and increased capitalist approaches to welfare programmes, economic development has taken precedence over welfare, social justice, and reducing inequalities in all societies. Neoliberal enterprises have forced governments to reduce investments in social security, health, and other welfare activities, shifting the focus towards opening up trade, removing trade barriers, and reducing export tariffs and internal subsidies to farmers and local businesses. This has had a devastating impact on the lives and livelihoods of people already on the margins, who live and work in precarity. Their vulnerabilities have only increased, and whatever social security the governments had been providing has been reduced or eliminated, leaving them to fend for themselves. The welfare state is seen as a problem in the neoliberal framework, as the ideology and politics emphasise individual enterprise and the dismantling of protective mechanisms for people and countries (Harvey 2010). Rather than alleviating poverty and deprivation, there has been an increase in inequalities both within and between countries. In this context, social work practitioners and organisations are once again at a crossroads, seeking new ways to approach welfare.

Spolander et al. (2014), in a six-country comparative study covering Finland, India, South Africa, Italy, England, and Russia, highlight structural shifts in social work practice and the double jeopardy confronting practitioners. Non-governmental organisations (NGOs), where many social work practitioners are employed, are partnering with governments to deliver services to welfare recipients, aiming to encourage outreach and the democratisation of services. However, they are simultaneously becoming ensnared in regulatory systems and reliant on government funding for their survival.

The focus of social work practice is shifting towards managerial approaches and specialisation in working with specific client groups (e.g. children), resulting in a loss of emphasis on community work, group interventions, casework with empowerment approaches, and critical pedagogy.

Meanwhile, philanthropic organisations are increasingly taking over welfare work from governments. These organisations often act as arms of companies and corporations that may violate the very rights of those they seek to assist, leading to agenda-driven initiatives. This situation leaves little room for social work practitioners to operate in a space that enables them to advocate for the values of social justice, equality, and the rights of the people (Pedicini 2011).

4.2.3 Scope of Transnational Social Work Practice

The Global Agenda for Social Work (IASSW 2024) and the SDGs provide opportunities for transnational collaboration, discussion, debate, and action on common agendas for social work practice and research. With the increasing influence of global players—such as multinational companies, the World Trade Organisation (WTO), and the International Monetary Fund (IMF)—on national government policies and programmes, social workers must adjust their practices and interventions both locally and globally. This shift necessitates the development of new skills and theoretical approaches to practice. Transnational social workers become the harbingers of a new global political space, engaging in discussions about inequalities and strategising on how to address these issues both within and among countries. Through transnational social work, the focus expands from the needs of those who are excluded to the mechanisms of exclusion itself, challenging the traditional notion of a defined 'inside' and 'outside' (such as social workers' focus solely on citizens). Transnational social work highlights inequality, poverty, and demands for global justice as interconnected realities that affect everyone's everyday life. This perspective adds a cross-national dimension to the pursuit of equality, underscoring the essential relationships between the local and the global (Schwarzer 2016).

4.3 Methodology

To examine social work practices and interventions in Germany, South Africa, India, and Malaysia, the study employed a qualitative design. Sixteen in-depth interviews and four focus group discussions (FGDs) were conducted with social work practitioners from these countries to gain insights into their professional experiences and realities. The tools for data collection included interview guides and FGDs. Of the sixteen interviews, six were face-to-face, while the remaining ten were held virtually. Similarly, two of the four FGDs were conducted in person, and the other two were held online to accommodate participants located in different countries. Participants were selected using purposive sampling, with the inclusion criterion requiring a minimum of five years of professional experience in the field. The practitioners represented a range of sectors and brought diverse expertise in areas such as criminal justice social work, health and clinical practice, child protection, and human rights advocacy. They were identified through participants involved in an international project (IWAAC 10), in which the authors have been actively engaged.

Out of sixteen social work practitioners, six were from Germany, four from South Africa, and three each from India and Malaysia, all of whom were interviewed individually. In Germany, two FGDs were conducted with social work practitioners who specialised in transnational social work programmes (undergraduate level), while the other two FGDs involved social work practitioners enrolled in the undergraduate social work programme at a university in Germany.

When contacting research participants, the purpose of the study was clearly communicated, and informed consent was obtained. Both in-person and virtual meetings were scheduled at times and locations convenient for the participants. They were informed of the steps taken to ensure confidentiality and anonymity throughout the research process. Verbal consent was obtained from all participants prior to their involvement in the study. With their permission, all interviews and FGDs were audio-recorded. The interviews, conducted in English, were subsequently transcribed for analysis.

A thematic analysis approach was employed to identify recurring patterns and themes within the participants' narratives across the four countries. However, due to the limited sample size, it was not possible to include social workers from all areas of practice. As a result, the insights presented are specific to the participants' respective fields and professional experiences.

4.4 Thematic Analysis

4.4.1 Diverse Roles and Context of Social Work Practitioners in the Four Countries

The findings revealed the variety and diversity of the work that social work practitioners are undertaking in the four countries. Germany has comprehensive social welfare programmes and provides strong social security support to its citizens. Social work practitioners in Germany primarily work within the welfare state to deliver these services to clients, mainly youth, children, and women. At the same time, they are now addressing the needs of a large population of internal and international migrants, which requires them to adopt new methodologies and processes to promote safe integration into the system.

In contrast, social work practitioners in the other three countries (Malaysia, South Africa, and India) operate within state systems across a variety of settings, such as health care, school social work, child protection programmes, and criminal justice systems, as well as with voluntary and NGOs that play a significant role in providing welfare services to clients. Consequently, practitioners in these three countries engage in collaborative processes, working with youth, communities, and collectives of women, among others. Their work thus combines engagement with state systems and direct interaction with people in communities. As a result, social work practitioners in these countries face a different set of challenges that arise while negotiating with state services and programmes for the public. Each country has its specific issues and social contexts, and social work interventions are organised accordingly.

In Germany, social interventions are more aligned with a person-centred approach, focusing on the needs and problems of clients, and casework as a method is predominantly used. Clara, a social work practitioner in Germany, questioned this particular approach during our conversation. She shared,

> "If we say, hey, poor people or under-supported individuals, do we really change the cause of why they are under-supported by our social work? Do we really address the grassroots, or do we just do social work? We can say we do something, but actually we do not change anything".

She reflected on emphasising the structures or systems and engaging with those that create vulnerability for individuals. The importance of working with systems or structures has been recognised by social workers in South Africa and India. Megan, a social worker from South Africa, highlighted how crucial this is for social work practice:

> "And so, to lessen the effects of the caseloads that we get, we need to do services, service rendering, preventative programs and put preventative measures in place. And that can only take place through engagement with group work and community services".

The preventive programmes mentioned here involve identifying the root of the problem, which derives from structures or systems, and engaging with them. The methods employed include group work and community work, which are largely used by social workers in three countries: South Africa, India, and Malaysia. However, in recent years, social work practice in Malaysia appears to have narrowed its analysis to individual and family levels, neglecting the community-based approach due to a shortage of staff and resources (Jayasooria 2016). The community work is extensively utilised by social workers in India and South Africa for their rights-based approach. Often, awareness building and information dissemination about individuals' rights are conducted at the community level. The community itself is considered an equal partner in developing interventions for the issues they face. The challenges encountered by the community, which are local in nature, require interventions that evolve from the unique needs, context, and resources of the community.

Aditi (India), a social worker who works with informal women workers, shared how their interventions are rooted in a community approach,

> "Majority of what we do is organising informal women in their communities. So, there is leadership development. The grassroots workers go into the community, and they identify leaders and they identify members who are informal workers. And they tell them about, you know, what their rights are, how they can connect them to different government schemes. And then again, from the community itself, they democratically elect their leaders so that each area has a leader, so to speak. Then, the leaders report to our grassroots community organisers".

All the social workers shared that their work includes networking and collaboration with other stakeholders. These collaborations create a space for discussions and engagement around issues and interventions. This is particularly relevant in the relationship between the state and the social workers employed in both state departments and voluntary organisations. It offers policymakers valuable insights from the field, as social workers are directly involved with the people they serve. Furthermore, certain issues can be highlighted to the state to develop appropriate interventions. Ultimately, these collaborations can lead to social workers becoming integral to the system. Although such collaborations exist in Germany, participants noted that the opportunity to provide feedback to the state is often limited. Anna (Germany), who provides migration counselling to migrant adults, reported that the state collects data on their work through questionnaires. However, she pointed out what is missing:

> "Documentation is proper. However, it doesn't say enough in statistics about their everyday life, about their everyday struggle, especially with authorities. Questions quite often have not that much to do with their everyday life. And if the question isn't asked in the statistics, it won't be counted".

The top-down approach often ignores ground realities, ultimately leading to exclusion and inequality. When interventions are predominantly managed by the state, there is less flexibility and limited scope for innovation. In spaces independent of the state, social workers have greater opportunities to design and formulate social interventions. This independence has been widely embraced by social workers in India and South Africa.

Nikitha (India), a social worker who provides support for children, explained her interventions,

> "We work with these family and strengthen these families so that the child doesn't have to get separated from the family. We link these families with social services schemes, with government or non-government services out there, with their Social Security documents, and we also provide certain direct financial support till the time the schemes come through, then we also provide different kind of sessions like financial literacy or certain life skills programme and certain parenting programme".

Even though space for advocacy is very limited in Malaysia due to the lack of recognition of the social work profession, social workers in the country are still able to carry out advocacy efforts, as reflected in the words of Alia, who works with binational families:

> "We definitely focus more on advocacy. I need to document more of this. We definitely need more proof, need more data and from all of that data, all of those live realities. We need to engage with the human resource ministry and Ministry of law, which we need to focus more on after this because most of our resources before this focus on citizenship issues. But we definitely have plans to focus more on this, to engage with the ministry, and to engage with members of Parliament so that this issue will get traction and then hopefully help trigger policy reform in this matter".

The nature of methods and interventions employed by social workers in the four countries varied according to the social problems and the nature of the state.

4.4.2 Experiences of Social Work Practitioners and Understanding of Inequalities

In the process of capturing the understanding of inequality among social workers, the core components of SDG 10 (inequality within and among countries) were explored, specifically poverty, migration, and discrimination. Although all the social workers acknowledged economic resources as one aspect of poverty, they identified additional factors to consider. For social workers in Germany, this included a lack of access to state benefits, while for social workers in India, it was a deficiency in social security and social capital. In South Africa, there was a lack of access to basic services, whereas social workers in Malaysia cited inadequate income and limited access to basic services.

Across the four countries, a migrant background was associated with disadvantages, making this group a prominent client demographic for social workers. For instance, in Germany, the language used for migration counselling is limited and depends on the social worker's knowledge of other languages. There is no mandatory requirement that ensures social workers with proficiency in languages other than German are employed by relevant organisations. As a result, a group of migrants who are not proficient in the language used for migration counselling is excluded from accessing these services. Participants noted that language is a significant barrier to accessing state-provided services in Germany. Lena, a social worker at shelter homes for the homeless population, shared the consequences of being a migrant.

> "For example, there are also elderly people here, and they are seriously ill. The social welfare office told me they could not do anything and that they had to go back to their country. Or if they stay here, they need to sleep in the street. because maybe they always worked here, but they worked in a black way, and they got robbed. I mean, a lot of people come here, they work, and they don't get paid. I mean, they're disadvantaged in a lot of ways because of the

language, because of nationality, because of their experience or whatever. And, then, maybe they get ill, and they don't have insurance, and it's difficult".

In India, migrants often lack the necessary documents to receive benefits, and social workers aim to address this as a first step. Similarly, in South Africa, social workers noted that the lack of documentation among asylum seekers leads to disadvantages. Often, these disadvantages stemming from migrant backgrounds result in discrimination and exclusion. In Malaysia, Mithra expressed concern that the state policies and programmes often fail to reach their intended beneficiaries, resulting in further exclusion. Amir also highlighted implementation challenges in relation to inclusive development, noting:

"But we have a new economic policy that we developed in 1971. Basically, the new economic policy [NEP] is to restructure the socio-economic policy for the poor and also to minimize the level of poverty in the country and also to achieve national unity, harmony and integrity. So, I think the policy is good. I think looking at the blueprints of this NEP, it looks good, but just like everything else, when it comes to implementation, there's a big problem, problems of governance, corruption and so on".

When asked about discrimination, Mosi (South Africa) shared that,

"Given our history, we take discrimination seriously, but when it comes to exclusion, people may be excluded. And it is, again, that urban-rural divide. It is the urban-rural divide that makes people to be excluded. Someone who lives in an area where a hospital is 350 km is excluded from healthcare services. Where there are no schools, in mining communities, where the rich mining companies have been taking resources from those communities, but they have not built a single school. That community continues to be impoverished, so they will be excluded because of their geographical location".

In India, social workers acknowledge the intersectional nature of discrimination arising from the overlapping influences of class, caste, religion, gender, and other social factors. Aditi (India) illustrated this complexity through the example of informal women workers:

"I think the intersectionality that a woman in this scenario faces is due to her gender, her livelihood, and her status of work, which is informal work. So when all of that plays out for her and the kind of burden that she then has, which is, I would say, a triple burden, because even in times of crisis, there are generally women who have to kind of absorb the shock".

The narratives clearly reveal that components of inequality—such as poverty, migration, and discrimination—are deeply interconnected. Addressing these issues in isolation is unlikely to produce meaningful outcomes; instead, an intersectional approach is essential for social workers to respond effectively.

4.4.3 Understanding of the Sustainable Development Goals: Local Relevance to the Global Agenda

The social workers in Germany demonstrated relatively low awareness of the SDGs compared to their counterparts in the other three countries. When asked about the SDGs, Anna from Germany remarked, "*Yeah, this is way too far away*". She observed that awareness exists up to the mid-level, but is not effectively communicated to frontline social workers. Similarly, Mosi from South Africa shared that,

"My general view will be, in South Africa, SDG is still very much a scholarly document that happens up there, you know, that we teach at university".

According to Zethe (South Africa), there is little to no discussion around the SDGs, which she described dismissively as "*It is just about numbers*". Amir from Malaysia expressed that,

"I think many of the social workers probably are not well versed with regard to SDG. I think maybe, I would say the whole country is perhaps not that well versed, except those very much working directly looking at the development, the future direction of the country. Maybe some politicians, some of them are aware. But in general, I would say there is a need to really, really talk about this SDG, you know, for social workers, not only for social workers, but also for other professions that are related or responsible in promoting SDG".

Amir (Malaysia) further noted that, despite teaching at various universities, he has not observed the SDGs being incorporated into social work curricula. He emphasised the importance of integrating the SDGs into academic programmes as a pathway to embedding them in professional practice. Social workers from Germany, South Africa, and Malaysia consistently stressed the need to raise awareness of the SDGs as a critical first step towards their integration into social work. They underscored the importance of extending this awareness to lower levels of departments and frontline social workers. Nikitha (India) also expressed concern about the practical adoption of the SDGs in everyday social work practice:

> "I feel like we have homegrown models and we've been doing well and we will figure out solutions to our complexity and issues. Your SDG's didn't necessarily address, you know, caste dynamics or whatever, all of that, our realities. But these things are already happening, but just because they were able to articulate it in a certain manner and put it in 1234, they're now mandating that we have to cover all of that … I think it needs to be more contextualized, even if it was to insist, on the country's reality. And even in the country, it's so diverse. If one state has the problem of female feticide, then another has the problem of child marriage, and another has the problem of so many other things, so it has to be contextualised is what I think. So translating it into our work is difficult. We cannot see the organicity, how it's organically coming down because what we do is way more complex than what they say".

Nikitha (India) emphasised the importance of contextualising the SDGs or decolonising them to better align with the realities of developing countries such as India. The local context of each country varies significantly, and often, interventions require a local perspective. Localising the SDGs will help address the unique needs of populations within their socio-cultural contexts and recognise them as stakeholders in the process. The majority of social workers proposed collaboration and networking as essential methods for integrating the SDGs into social work practice. They advocated for engagement not only within national boundaries but also across them, supporting a transnational approach that fosters shared learning and collective action.

4.4.4 Scope for Transnational Social Work and Realisation of the Sustainable Development Goals

The need for a transnational approach in social work to achieve the SDGs emerged clearly from the FGDs conducted with social work practitioners engaged in transnational social work programmes. These professionals recognised the global interconnectedness of social issues and viewed the SDGs as a valuable framework for organising their work, regardless of the country in which they are based. More importantly, the SDGs offered a means to track progress and facilitate cross-country comparisons. Social workers practising in developing countries also noted that the SDGs are particularly relevant for securing funding for interventions. Joy (Germany), a transnational social worker, elaborated,

> "And especially when I am in another country now, I have noticed that since the goals are there, there is always at least a name where we can ask for when it comes to funding in these countries. It's kind of a hype thing. Like everybody says, okay, we are trying to develop this goal or supporting sustainable development goal. And in the name of sustainable developmental goals or particular goals, I think more funds are available focused on a particular topic available for development in a particular area".

Transnational social work exchanges enable social workers to be culturally sensitive to the realities and social norms of different contexts, while also exposing them to indigenous social work practices and approaches. The lack of resources in developing countries has been cited by transnational social workers as a challenge in achieving their goals. However, this understanding, which largely stems from transnational exposure, is relevant when working within the framework of the SDGs. Joy (Germany) expressed his feelings about this,

"But at the same time, the conflict I always have in mind is the lack of resources. What most countries go through; even if the social workers know there is an existing problem, they can't touch the problem or do anything for the problem due to the lack of resources. It's just understanding. If anything can be done on it, who knows? But at least there is awareness of the lack of resources for the world out there compared to what they have in Germany and an understanding of them rather than criticising them".

4.5 Discussion

The previous section presents the findings of a qualitative research study on the experiences of social work practitioners in their chosen sectoral areas across four countries: Germany, Malaysia, South Africa, and India. The findings reveal a rich diversity in social work practices in these countries, significantly shaped by their differing welfare models. The substantial influence of Esping-Andersen's (1990) typology of welfare states on social work is well-documented in the literature, evident in the methods and practices adopted in the four countries.

The strong presence of the welfare state in Germany enables social workers to adopt a person-centred approach, predominantly using casework. However, this same presence often limits innovation and advocacy among social workers. In contrast, the lack of a robust welfare state in the other three countries encourages social workers to adopt a more community-based and rights-based approach, alongside opportunities to design and formulate social interventions. The limited capacity to provide feedback to the state and the top-down approach in certain fields of practice, as cited by social workers in Germany, is a common phenomenon. Often, social workers are marginalised concerning macroeconomic decisions and must witness the negative consequences of these decisions, which promote social exclusion and deepen existing social inequalities (Muleya 2020). The integrated approach, which requires both micro- and macro-level interventions, has been a significant topic in social work literature (Kaloga and Reno 2022). When space is limited for social workers to operate on a macro level, particularly in engaging with systems, adopting an integrated approach becomes challenging.

The interconnectedness of poverty, migration, and discrimination, as recognised through SDG 10, is evident in the narratives of social workers across the four countries. The participants' operationalisation of poverty as a lack of access to state benefits, social security, and basic services reveals the impact of the neoliberal framework, which prioritises the removal of protective mechanisms for individuals and nations (Harvey 2010). The discrimination and exclusion faced by specific communities, such as migrant communities, in each country necessitate a 'glocal' mindset—one that combines global awareness with a focus on local solutions (Ghanem et al. 2020). The SDGs, as a well-articulated framework, enable social work to simultaneously engage with globalised influences on local contexts while recognising the primacy of those local realities (Kaloga and Reno 2022). The diversity in responses to social issues such as poverty, migration, and discrimination illustrates how social work adapts to varying welfare regimes and social conditions.

The low awareness of the SDGs among social workers calls for increased attention to enhance understanding. With a clearer awareness and understanding of the SDGs among social workers, the integration of these goals into practice will be expanded. As pointed out by the participant, the need to contextualise or decolonise the SDGs to fit into developing countries has been a topic of discussion in SDG literature. The SDGs have been criticised for oversimplifying complex issues, offering empty symbolism, and perpetuating Western colonialism (Schultz 2020). The findings highlighted the necessity of localising the SDGs to meet the unique needs of the population. The United Nations Development Programme (UNDP) has described the process of SDG localisation as the 'recognition and empowerment of vital local actors (particularly local governments, populations, businesses, small and medium enterprises, and community-based organisations) in developing, implementing and reviewing SDG-related measures in cities and communities' (UNDP 2015:4). These local actors, with whom social workers work closely, can support the achievement of the SDGs through bottom-up action (UNDP 2016). This concept is also closely related to indigenisation, which Barise (2005) describes as a 'bottom-up' process that develops approaches rooted in the local context and adjusts mainstream social work accordingly.

Participants emphasised the importance of collaboration and networking across borders, advocating for a transnational approach to social work. Social workers engaged in transnational programs strongly supported this perspective, asserting that such an approach is essential for effectively advancing the SDGs. The complexities that arise from the transnational context can only be addressed through transnational social work. The SDGs provide a framework for addressing globally interconnected issues and also serve as a catalyst for securing funding in developing countries. Moreover, a transnational approach offers social workers exposure to Indigenous social work practices, fostering opportunities to learn from diverse experiences and perspectives across national contexts (Schwarzer 2016). The potential of transnational social to meaningfully contribute to SDG 10, which focuses on reducing inequality within and among countries, should not be underestimated.

4.6 Conclusion

A closer examination of the SDGs reveals a strong alignment between their objectives and the core principles of social work. In an era shaped by globalisation and neoliberalism, where social issues increasingly transcend borders, the role of contextually informed social workers in advancing sustainable development is more critical than ever. This chapter has explored the diverse nature of social work practice in four countries—Germany, South Africa, India, and Malaysia—using a qualitative approach to examine social workers' understanding of the SDGs. The findings indicate that social workers are addressing transnational social problems through varied methods and practices, shaped by the specific welfare regimes of their respective countries. Transnational social work, which has the potential to cross boundaries rather than accept them (Schwarzer 2016), emerges as a promising approach to advancing the sustainable development goals. However, the relevance of indigenous knowledge and practice must not be overlooked in the adoption of transnational social work.

References

Barise A (2005) Social work with Muslims: insights from the teachings of Islam. Crit Soc Work 6(2)

Bodhi SR (2011) Professional social work education in India, a critical view from the periphery. Indian J Soc Work 72(2):230

Bodhi SR (2020) Tribes and state policy in India: revisiting governing principles from a decolonial social work perspective. Br J Soc Work 50(8):2372–2388. https://doi.org/10.1093/bjsw/bcaa186

Chetty D (1999) Social work in South Africa: historical antecedents and current challenges. Eur J Soc Work 2(1):67–76. https://doi.org/10.1080/13691459908413806

Esping-Andersen G (1990) Three worlds of welfare capitalism. Policy Press, Bristol

Ghanem A, Nazzal D, Hussein F et al (2020) Localisation of social work in Arab countries: Basic paper. https://ifas.fhws.de/fileadmin/redaktion-loswac/Basic_Paper_-_Localisation_of_Social_Work_in_Arab_Countries_DAAD_LOSWAC.pdf. Accessed 23 July 2021

Gore P, Gore M (1977) Social work education in India. Tata Institute of Social Sciences, Bombay

Gore MS (1997) A historical perspective of the social work profession. Indian J Soc Work 58:442–455

Gray et al (2013) Decolonizing social work. Routledge, London

Hammerschmidt P, Weber S, Seidenstücker B (2017) Soziale Arbeit—die Geschichte. Barbara Budrich, Opladen

Harvey D (2010) A brief history of neoliberalism. Oxford University Press, Oxford

Hatta AZ et al (2014) Professional and alternative social workers: a case study of Malaysia. Asian Soc Work Policy Rev 8:138–155. Wiley Publishing Asia Pty Ltd.

IASSW (2024) Global definition of social work https://www.iassw-aiets.org/global-definition-of-social-work-review-of-the-global-definition/. Accessed 18 Apr 2025

Jayasooria D (2016) Sustainable development goals and social work: opportunities and challenges for social work practice in Malaysia. J Hum Rights Soc Work 1:19–29. https://doi.org/10.1007/s41134-016-0007-y

Kaloga M, Reno R (2022) SDGs and systems science: teaching beyond the micro-macro divide in social work education. Soc Work Educ 41(7):1489–1506. https://doi.org/10.1080/02615479.2022.2109622

Laging M, Schaffer P, Lorenz M (2021) Social work and social work education in Germany: development and challenges in a scientific and practice-based profession and its education. In: Laging M, Žganec N (eds) Social work education in Europe: European social work education and practice. Springer Nature, Cham, pp 105–126
Mishra A (2021) Professionalization of social work in colonial India: glancing at the history of social work before 1936. Int Soc Work 1–12. Sage Publications
Mohammad AA, Uddin S (2020) Starting the indigenisation process of social work practice in Malaysia. Soc Work Soc 18(1):1–14
Muleya E (2020) Developmental social work and sustainable development goals in South Africa: opportunities and challenges. Int J Community Soc Dev 2(4):470–486. Sage Publications. https://doi.org/10.1177/2516602620975226
Pedicini J (2011) The two Indian welfare systems: state & corporate responses. In: Barnett papers in social research. Oxford University Press, Oxford
Schultz R (2020) Closing the gap and the sustainable development goals: listening to aboriginal and Torres Strait islander people. Aust N Z J Public Health 44(1):11–13. https://doi.org/10.1111/1753-6405.12958
Schwarzer B (2016) Transnational social work: an introduction. In: Schwarzer B, Kämmerer-Rütten U, Schleyer-Lindenmann A, Wang Y (eds) Transnational social work and social welfare: challenges for the social work profession. Routledge, New York, pp 1–306
Singh S, Gumz E, Crawley B (2011) Predicting India's future: does it justify the exportation of US social work education? Soc Work Educ 30(7):861–873
Smith L (2014) Historiography of south African social work: challenging dominant discourses. Soc Work Maatskaplike Werk 50(3):304–331
Spolander G, Engelbrecht L, Martin L et al (2014) The implications of neoliberalism for social work: reflections from a six country international research collaboration. Int Soc Work 57(4):301–312
United Nations Development Programme (UNDP) (2015) Capacity development: a UNDP primer. https://www.undp.org/content/undp/en/home/librarypage/capacity-building/capacitydevelopment-a-undp-primer.html. Accessed 18 Mar 2025
United Nations Development Programme (UNDP) (2016) Roadmap for localizing the SDGs: implementation and monitoring at sub-national level. Global Taskforce of Local and Regional Governments; UN-Habitat; UNDP, Washington, DC
Wendt WR (2017) Geschichte der Sozialen Arbeit 1: Die Gesellschaft vor der sozialen Frage 1750 bis 1900. Springer VS, Wiesbaden
Willmore B (1985) Training for flexibility in choice of method in social work practice: case study. Dissertation, University of Zimbabwe

Using Sustainability Research to Address Social Inequalities

5

Adeel Sambo, Rinie Schenck, and Nicolette V. Roman

Abstract

Research is conducted across a wide range of disciplines and serves as a cornerstone for the generation and dissemination of valuable information. This chapter introduces sustainability research with a particular focus on addressing reduced inequalities—an essential dimension in the pursuit of a more just and equitable world. As natural ecosystems and living conditions face increasing degradation on a global scale, sustainability research is becoming increasingly important both nationally and internationally. This growth has accelerated particularly since the introduction of the Sustainable Development Goals (SDGs). Research aids in the creation of knowledge by meticulously converting collected information into new knowledge. The disseminated knowledge contributes to the progression of societies and fosters a deeper understanding of how we perceive the world. This, in turn, enhances overall development. This chapter emphasises the importance of research in addressing SDG 10 by focusing on social intervention. To highlight the relevance of research, the authors explore several targets encompassed within SDG 10, including the investigation of trends in inequality and the definition of inequality itself. This research assists in the formulation of policies relevant to vulnerable communities, thereby raising awareness and tracking progress on improvements through the allocation of resources. The authors aim to connect these targets by scrutinising the analytical tools used to address social inequalities across countries and applying intersectionality as a framework.

Keywords

Sustainability research · Sustainable development goals (SDGs) · Reduced inequalities · Social discrimination · Social development

A. Sambo (✉)
Cape Peninsula University of Technology, Bellville, South Africa
e-mail: sambom@cput.ac.za

R. Schenck · N. V. Roman
Centre for Interdisciplinary Studies of Children, Families and Society, University of Western Cape, Bellville, South Africa
e-mail: cshenck@uwc.ac.za

N. V. Roman
e-mail: nroman@uwc.ac.za

5.1 Introduction

> "As long as poverty, injustice, and gross inequality persist in our world, none of us can truly rest"
> Nelson Mandela (2005).

Over the past thirty years, significant progress has been made in the discourse on sustainable development. While developed nations have generally advanced further, many emerging countries have

N. V. Roman et al. (eds.), *Social Inequalities Within and Among Countries*, Sustainable Development Goals Series, https://doi.org/10.1007/978-3-032-08741-6_5

also increasingly recognised the importance of pursuing sustainability. The United Nations (UN) continues to play a central role in supporting countries as they address the complex challenges of sustainable development (Salvia et al. 2019). In September 2015, the UN and its 193 member states adopted the 17 Sustainable Development Goals (SDGs), committing to their full implementation and achievement by 2030 (Biermann et al. 2017). In academia, research related to the SDGs is emerging across a range of disciplines, including business and management. Some of this research explores and demonstrates the energising effects of adopting an SDG framework to guide organisational policy and practice (Bebbington and Unerman 2015). Moreover, research serves as a platform for scholars to engage ethically with local communities, helping to identify and address the challenges faced by vulnerable individuals (Kluczyńska et al. 2024).

To achieve the SDGs, research, innovation, and sustainable education are essential. Research not only serves as the foundation for progress and development across various fields but also shapes our understanding of the world. It plays a vital role in addressing complex challenges and fostering creativity and innovation (Islam 2023). The implementation of such research activities will require substantial investment from both public and private sectors. In fact, the effective application of research findings will depend on an integrated approach involving educational institutions, national, local, and municipal authorities, as well as commercial enterprises, civil society organisations, and international bodies (Salvia et al. 2019). Sustainable development interventions have made significant contributions to addressing environmental and socioeconomic challenges. While substantial effort has been invested in designing and implementing such initiatives, comparatively less attention has been given to ensuring that the research underpinning these efforts is conducted in a sustainable way (Bebbington and Unerman 2015). To understand and address pressing sustainability-related challenges, researchers in this field may focus on developing appropriate sustainability indicators, validating existing ones, and using pertinent approaches (Horsley et al. 2015). The enhancement of sustainable growth and healthy lifestyles will strengthen research and inform those responsible for decision-making and policy development (Filho et al. 2017).

The core objective of this chapter is to establish a connection between SDG 10 and sustainability research. To achieve this aim, the authors present sustainability research as an overarching theoretical perspective through which research can be integrated with socio-economic and environmental inequalities. Sustainability research spans a wide range of topics and draws from various academic disciplines. These include environmental challenges such as food shortages, climate change, and biodiversity loss, as well as social and economic issues such as income inequality, poverty, and political division (Liu et al. 2021). This chapter will first outline some background information on the SDGs, followed by a discussion of sustainability research. It then provides a brief overview of SDG 10. Before concluding, the chapter illustrates how sustainability research can be applied to address factors relating to SDG 10.

5.2 Background Information on the SDGs

Sustainability, whether regarded as a theory, method, or practice, involves making effective use of accessible resources to ensure there will be enough to meet the demands of both the present and future generations. Another definition of sustainability is the ability to allocate resources responsibly to both economic and non-monetary endeavours in an attempt to achieve specific social, economic, and environmental goals (Peterson 2022).

In 1982, a commission was formed, chaired by Gro Harlem Brundtland, Prime Minister of Norway, with Mansour Khalid, Deputy Prime Minister of Sudan, serving as Vice Chairperson (World Commission on Environment and Development 1987). In 1987, the Brundtland Commission published a report titled 'Our Common Future'. The report explicitly outlined the characteristics of sustainable development, defining it as 'development that meets the needs of the present without compromising the ability of future generations to meet their own needs' (South Africa. Department of Environmental Affairs 2016).

In 2000, the UN launched the Millennium Development Goals (MDGs) to combat poverty and its many facets (UN 2015). The United Nations Millennium Declaration was adopted by the largest gathering of world leaders at the Millennium Summit in September 2000. It established a set of goals with a target date of 2015 and committed participating countries to a new global partnership aimed at reducing extreme poverty. This initiative became known as the Millennium Development Goals (MDGs) (Asian Development Bank 2009).

In the final Millennium Development Goals report (MDGR), Minister TJ Radebe stated that the MDGs were a perfect fit for South Africa, as the fundamental rights they upheld were already enshrined in the Constitution of South Africa (REF). Moreover, the MDGs complemented the development goals articulated in the South African Freedom Charter, which has been supported by several governments since 1994.

In 2015, the UN introduced the 17 SDGs. For many years, countries have struggled with the triple challenges of poverty, unemployment, and inequality (UN 2016). The report further notes that, despite progress in reducing poverty, one in eight individuals globally continues to live in extreme poverty. At the 2015 summit, India's Prime Minister affirmed the country's full support for the SDGs. He emphasised that 'much of India's development agenda is mirrored in the Sustainable Development Goals' and highlighted the stark contrast between the world's immense wealth and the persistent, unimaginable deprivation that still exists. He further reaffirmed the importance of the SDGs, noting that India has made significant progress in advancing the SDG agenda through notable achievements (National Institution for Transforming India [NITI] 2018).

Furthermore, Malaysia has demonstrated its commitment to the SDGs by reducing poverty by 90% between 1990 and 2014. Over the past 50 years, the country has maintained an average annual economic growth rate of 6.2%, contributing to a 54% increase in the female labour force (Economic Planning Unit 2017). Germany, on the other hand, has achieved 25 of the 130 SDG targets and is projected to reach an additional 12 targets (Organisation for Economic Co-operation and Development [OECD] 2022).

Sustainable development is anticipated to yield long-term benefits for both the environment and all people at the socioeconomic level (Peterson 2022). Within the social sciences, the concept of sustainable progress is grounded in the rejection of poverty, unemployment, gender-based discrimination that undermines women's rights, and the growing disparity in wealth. In this context, the social component serves as a foundation for the concept's sustainable development through social justice, with the most crucial elements being the achievement of equitable conditions between present and future generations (Al-Taai 2021). New research should seek ways to integrate the objectives of sustainable development into the government's existing agenda to provide national governments with local solutions to common issues. Subsequent studies ought to investigate how sustainable development and sustainability objectives might enhance people's quality of life and economic prosperity (Peterson 2022). Liu et al. (2021) assert that as natural environments and ecosystems that support life are being damaged on a large scale, sustainability research is becoming increasingly important at both national and international levels. This expansion has been particularly rapid since the recently established SDGs were introduced.

In relation to SDG 10 (Reduced Inequalities), the goal mandates the reduction of financial disparities as well as inequalities based on age, sex, disability, caste, race, ethnicity, religion, and opportunity—both within and among nations. Economic inclusion refers to access to financial services, comprehensive education, and meaningful employment; social inclusion involves equitable access to social resources and full participation in community life; and political inclusion refers to the ability to vote both actively and passively (GermanWatch and CAN Tanzania 2017). Küfeoğlu (2022) asserts that SDG 10 is a significant factor in fostering a sustainable global environment. Although substantial progress has been made since the turn of the millennium in lifting individuals out of poverty, considerable disparities in wealth and income persist. Groups such as rural populations (including farmworkers), women, people with disabilities, and youth are often situated at the lowest levels of the income tier (World Bank Group 2019). One of the global challenges is that, although inequality between nations is declining, inequality within countries continues to rise (GermanWatch and CAN Tanzania 2017). In addressing this issue, Küfeoğlu (2022) categorises inequality into the following groups:

- Economic Disparity: Refers to differences in income, consumption, and wealth among people in societies.
- Social Discrimination: Arises from distinctions in social status and positions, causing economic disparities.
- Political Inequality: Reflects unequal outcomes resulting from decisions made by political authorities, which may lead to the exclusion of certain groups from political participation.
- Environmental Dissimilarity: Evident in the unequal distribution of ecosystem benefits—such as access to clean water, parks, and land—across population groups.

Social, economic, and environmental inequalities are closely linked to changes in individuals' social status and roles within society. One common measure of income inequality is the Gini coefficient, which ranges from 0 to 1. A value closer to 1 indicates a higher level of inequality, while a value closer to 0 reflects a more equitable distribution of income (Makgetla 2020). The wealthiest individuals residing in the richest countries are 134 times wealthier than those living in the poorest countries (GermanWatch and CAN Tanzania 2017). We are confronted by an unparalleled extreme inequality crisis, as stated by Oxfam (Development Finance International 2023:4):

> We are living through an unprecedented moment of multiple crises. Tens of millions more people are facing hunger. Hundreds of millions more face impossible rises in the cost of basic goods or heating their homes. Poverty has increased for the first time in 25 years. At the same time, these multiple crises all have winners. The very richest have become dramatically richer and corporate profits have hit record highs, driving an explosion of inequality.

5.3 Sustainability Research

The term 'sustainability research' refers to studies that seek to balance resource use across environmental, social, and economic domains while addressing complex societal challenges. Change is necessary to move away from unsustainable practices that hinder long-term viability (de Camara et al. 2021). This approach ensures that present needs are met without compromising the ability of future generations to meet their own. Sustainability research often employs multidisciplinary and cross-disciplinary methods, emphasising the interconnectedness of various dimensions of sustainability rather than examining them separately (Miya et al. 2021). Becker (2012) argues that sustainability research cannot be fully recognised without the integration of both ethics and science. She further emphasises that excluding ethical considerations from sustainability research hinders its development. Sustainability ethics encompasses structural ethics and incorporates values rooted in community morals and behaviour aimed at improving the environment. Integrating sustainability ethics can serve as a foundation for promoting social, economic, and environmental well-being, positioning individuals and institutions as ambassadors for sustainable development (de Camara et al. 2021).

Some of the significant characteristics of sustainability research are discussed below:

1. **Multidisciplinary Cooperation**

Sustainability research entails integrating knowledge from several academic fields to address complex sustainability issues in a holistic manner.

2. **Long-Term Vision**

Sustainability research emphasises consideration for the needs of future generations by prioritising long-term impacts and sustainable solutions over short-term gains.

3. **Considering Perspective on Ethics**

Sustainability research should be conducted from an ethical perspective that considers social justice, economic inclusion, and environmental boundaries.

4. **Development for Sustainability**

Sustainability research aims to create innovative technologies, processes, and concepts that advance sustainability objectives, such as fostering social cohesion, lowering carbon emissions, and. Improving energy efficiency.

5. **Involvement of Stakeholders**

To ensure that research findings are relevant and applicable to real-world challenges, sustainability research often involves engagement with stakeholders, including communities, policymakers, and the business sector.

Each of the characteristics outlined above will be applied in the following section when discussing SDG 10. By focusing on key attributes, sustainability research seeks to contribute meaningfully to the achievement of the SDGs, established by the UN as a comprehensive and integrative framework for addressing global challenges. It is important to note, however, that there remains a limited body of research explicitly linking sustainability research to SDG 10.

Salas-Zapata et al. (2017) argue that, in the context of ecosystem protection, a new paradigm—more human-centred and grounded in increased awareness of sustainability—is needed. Researchers and individuals are encouraged to pose reflective questions such as, *what does this mean?* Miya et al. (2021) further ask, *what constitutes sustainability research activities, and how can sustainable practices be integrated into research projects? How should we begin?* To address these questions, they propose a set of practical guidelines designed to support community-wide engagement in sustainability research.

Ligozat et al. (2020) present a range of standards that encompass diverse topics, each with varying degrees of potential impact. Based on their framework, the following criteria are suggested:

Criterion 1: The acceptance of random selection
Begin your journey towards sustainability research by selecting one of the criteria below and committing to it. Some criteria may be simpler to follow than others, depending on your specific area of study or interests. Researchers should always be cognisant of their actions and recognise that introducing sustainability into their research needs to be done gradually, as it is a substantial initiative.

Criterion 2: Acquaint oneself with facts
The basis for sustained endeavours is knowledge. Reducing the negative effects that impact the SDGs can only be achieved if we are all aware of these effects. For instance, it is only possible to significantly reduce the carbon footprint of human activity if we have a thorough understanding of the specific impacts of various carbon-emitting activities. Educational programmes on how to protect the environment should be made available to researchers and others.

Criterion 3: Integrate sustainable techniques
Researchers should schedule their research carefully in advance, make effective use of their time, communicate regularly with colleagues, develop policies for sustainable practices, and provide incentives for sustainability within their groups or organisations. For example, instead of physically attending workshops or conferences, remote attendance could help combat pollution by limiting travel. Miya et al. (2021) observed during an e-conference that researchers were increasingly discussing ways to reduce the need for air travel to academic events, which often limits participation. They proposed that combining face-to-face and online interactions could enhance inclusivity by enabling broader participation among academics and researchers. Furthermore,

Miya et al. (2021) encourage scholars to critically reflect on their research practices, considering what could have been done differently to reduce environmental impact and promote sustainability.

Criterion 4: Share information, engage, and interact

A well-structured research design ensures replicability and minimises redundancy by building on existing knowledge. Providing detailed documentation allows other researchers to effectively replicate and validate the study. Innovation in research does not always stem from the use of new methodologies; it can also emerge from applying established methods to novel concepts. Salas-Zapata et al. (2017) highlight that the literature reveals a lack of multidisciplinary stakeholder involvement, decision-makers, problem-solvers, and systems-based methodologies in efforts to integrate sustainability into research. McManners (2019) further argues that while research is inherently valuable, research specifically focused on sustainability is particularly beneficial in generating new concepts and identifying emerging problems—ultimately informing changes in social, environmental, and economic policy.

Participatory Action Research (PAR) is a research approach in which researchers immerse themselves in the lives of the participants. PAR is a cyclical process through which participants identify an issue, develop an approach tailored to their context, implement their proposed solution, and gather evidence to evaluate its success. This iterative process is repeated until all parties are satisfied with the outcome (Ferrell 2014). Chukwu (2015) further notes that one of the main features of PAR is that it involves a series of self-reflective cycles, including planning a change, acting on it, observing its effects, and reflecting on them before repeating the sequence.

Researchers can make a substantial contribution to tackling sustainability concerns, reducing environmental impacts, and fostering ethical behaviour within the field of science by adhering to these standards and integrating sustainable practices into their research initiatives (Wuelser and Pohl 2016). Peterson (2022) claims that efforts to mitigate inequality and address climate change are not making significant progress. This could be improved if PAR is implemented to facilitate social interaction among individuals. PAR is a framework that encourages dialogue between practitioners, researchers, and other stakeholders (Goodyear 2013).

Criterion 5: Stimulate a bottom-up approach

Although this section advocates for a bottom-up approach, it does not suggest that the top-down approach should be disregarded. Kim et al. (2014) argue that operations strategy has traditionally been conceptualised as a top-down process of formulation and implementation, aligned with the broader corporate strategy. This perspective is grounded in the belief that manufacturing and operational decisions should support overarching organisational objectives. The top-down approach has gained substantial traction and has shaped empirical research on operations strategy. Conversely, Wilson and Irvine (2013), drawing on stakeholder theory, propose that a bottom-up approach—where individuals actively participate rather than passively receive information—is more likely to succeed than approaches that rely solely on top-down decision-making by management.

Recognising the value of community knowledge and action is essential, as is the shift toward a bottom-up approach that grounds policy development in locally appropriate practices. Bridging this gap requires giving due respect to Indigenous knowledge systems and local coping strategies (Shaw et al. 2008).

Community participation is paramount to ensuring awareness and adoption of sustainable activities. Developing members of the community will contribute to addressing issues that can be fostered within the community culture, providing further support to enhance sustainability. McManners (2019) emphasises that research focusing on sustainability should adopt an action research approach. In this approach, the researcher becomes a respondent seeking the most relevant solutions to the conundrum, rather than being guided by hypotheses or research questions.

Becker (2012) states that when designing sustainability research, the following four key components must be considered: a) amalgamation of various fields; b) combining humanities and sciences; c) localisation, addressing ambiguity and time; and d) acting to solve complications.

5.4 Accomplishing SDG 10 Through Sustainability Research

The tenth Sustainable Development Goal—SDG 10—seeks to reduce inequality within and among nations (UN 2016). Scholars can make a substantial contribution to eliminating social inequities and fostering inclusion by coordinating sustainability research practices with the objectives of SDG 10. It is important to understand that SDG 10 highlights how inequality restricts opportunities and can lead to societal discontent and violence, emphasising the need to eliminate disparity as a fundamental barrier to sustainable development. Coordinated action is needed to address resource distribution, income inequality, and disparities both within and between nations to achieve the goals of SDG 10. It is crucial to understand that SDG 10 highlights how inequality restricts opportunities and can lead to societal unrest and violence, positioning inequality as a significant barrier to sustainable development. Collaborative efforts are essential to address challenges related to resource availability, wage disparity, and inequality at both national and international levels to achieve the objectives of SDG 10 (UN 2015). Studies indicate that sustainability research, such as sustainable financial practices, effectively reduces income inequality, especially in settings with inadequate regulatory systems (Habib et al. 2024). Research acts as a driver for change by generating reliable data and identifying patterns and trends that can inform policies aimed at reducing inequality and promoting social inclusion (Knight et al. 2014). Moreover, by focusing on topics related to inequality, engaging with marginalised individuals, and supporting fair initiatives and processes, researchers can align their efforts with SDG 10 (Ghosh et al. 2023).

Social inequality stems from systemic issues deeply embedded in political, economic, and historical structures (United Nations Educational Scientific and Cultural Organization 2016). To effectively address these challenges, interventions must be informed by education and grounded in evidence. Research plays a critical role in identifying root causes, recurring patterns, and potential solutions, given its capacity to evaluate, interpret, and explain complex social problems (Dutonde 2018). SDG 10 comprises ten targets; however, this chapter focuses on four of them, namely:

Target 1—Reduce income inequalities.

Target 3—Ensure equal opportunities and end discrimination.

Target 4—Adopt fiscal and social policies to promote equality.

Target 7—Responsible and well managed migration policies.

Important functions of research in resolving social inequalities:

i. ***Addressing Inequality Trends (Target 1)***

Research plays a vital role in identifying various forms of inequality and, through data analysis, contributes to the development of appropriate measurement tools and policy recommendations (Stewart 2013). Ultimately, this supports the design of effective interventions aimed at addressing inequalities. Furthermore, research enhances the understanding of inequality as experienced by vulnerable populations and informal dwellers (Michael et al. 2019). Below are several examples in which research has proven successful as a tool for intervention in addressing social inequalities:

Community Health (Vaccarella et al. 2019)

Any disparities in health that are perceived as unjust or resulting from injustice are referred to as health inequities. The general term for discrepancies, variances, and inequalities in the health outcomes of individuals and groups is 'health inequality'. An evaluation of a health promotion programme designed to lessen socioeconomic disparities in cancer incidence was conducted. It became clear that actions taken to address the population, through media advertisements, may inadvertently enhance inequality in society. However, deeper and more drastic measures had a greater tendency to decrease inequality. Socioeconomic and political choices are the root causes of health inequities. In the United Kingdom (UK) and the United States of America (USA), health disparities significantly decreased during the 1920s and 1970s, when social welfare systems were established and wealth disparities diminished. Scotland's population is becoming healthier overall, and the death rate is also declining. Nonetheless, there remains an excessive number of untimely deaths in Scotland, and the differences in well-being and health between the richest and poorest individuals are growing (NHS Health Scotland 2016).

Economic Inequality (Doyle and Stiglitz 2014)

Severe disparities often hinder economic growth and threaten social stability and political equality. Among the various forms of inequality, inequality of opportunity is arguably the most damaging, as it both stems from and contributes to unequal outcomes. It also impedes economic development by creating inefficiencies, as large segments of the population are unable to realise their full potential. Countries with high levels of inequality tend to invest less in public goods such as infrastructure, technology, and education—key drivers of long-term economic growth and development. In contrast, reducing inequality offers clear social and economic benefits. When individuals perceive society as fair, social cohesion and mobility are strengthened, increasing the likelihood that more people will reach their potential and expanding public support for inclusive growth programmes.

Education Discrimination (Akala 2018)

In South Africa, the goal of gender equality in society, particularly in education, has not been achieved despite new legislation introduced after 1994. While some gender disparities have been addressed, educational institutions are primarily responsible for the persistence of gender prejudice ingrained in their systems.

ii. ***Recognising Groups That are Vulnerable (Target 3).***

Research can be used to identify populations with vulnerabilities that are subjected to social injustices (Patel et al. 2017). It also addresses the voices of those in precarious situations and provide assistance (Maringe 2014).

A report by the International Labour Organisation (ILO) and the OECD Development Centre highlights that two billion people are currently employed in informal jobs. This widespread reliance on informal employment presents significant risks and vulnerabilities, posing a major policy challenge—particularly in developing countries, where a large proportion of the population depends on the informal economy in various ways (OECD and ILO 2019).

Daniels et al. (2022) conducted a systematic review outlining the implementers, obstacles, and effective techniques for putting these interventions into practice, targeting vulnerable groups such as those involved in the criminal system and those living in unstable housing. The study highlights that for an intervention to be implemented successfully, there must be an atmosphere of support, as well as self-efficacy and confidence among both patients and providers. It also emphasises the importance of addressing stigma, patient needs, and the real-world concerns of staffing and intervention costs. The research focuses on a broad spectrum of interventions across various institutions, including coordinated medical systems and community-based programmes, encompassing technology-driven, peer-to-peer, and pharmacological approaches.

iii. ***Educating the Development of Policy (Target 4).***

The implicit assumption that reducing inequality is too formidable a challenge for social science research to address without major changes to the social policy landscape in the USA may partly explain the disconnect between much of the existing sociological research on disparities and the development of a robust, solutions-oriented knowledge base. Thus, a lot of sociological research on inequality concentrates on offering a narrative framework that logically suggests the societal advantages of the political realignment required to promote an activist and proactive social policy climate, whether overtly or implicitly (Diprete and Fox-Williams 2021). Research sponsors are increasingly requesting that their grantees address how research findings are incorporated into policy and decision-making processes. This expanding trend is a reaction to the necessity for context-sensitive, real-world data to solve the challenges that policymakers have in addressing intricate healthcare systems and health care administration inefficiencies (Erismann et al. 2021).

iv. ***Progress Monitoring (Target 7).***

Research monitors progress on inequalities through feedback provided by various stakeholders. Since the introduction of the SDGs, the UN has reported annually on this progress. Research information obtained from different countries allows for easy monitoring of advancements. This is the first time that the global development agenda has acknowledged migration as a key concern. The 2030 Agenda recognises that although migration can present new challenges, it can also contribute to development. The SDG objectives highlight aspects such as the necessity for increased anti-human trafficking efforts and the need to lower the overall cost of remittance transactions to improve development outcomes (Mosler Vidal and Laczko 2022).

In certain nations, the degree to which working circumstances are regulated by government agencies may be influenced by other elements, such as a lack of knowledge about the hazards involved or even a disregard for the well-being and security of migrant workers and those in unsafe jobs. Furthermore, temporary migrant labourers, who frequently have little legal or other protection, are essential to the agricultural industries of various nations (International Council for Science 2017).

The OECD and ILO clearly state that their goal is to present novel approaches to social policy that, when combined with other initiatives to increase labour income and productivity, can address the issue of informality in general, lessen the precariousness of individuals employed in the informal economy, and serve as cornerstones of equitable society, sustainable growth, and honest employment (OECD and ILO 2019).

To attain the targets of SDG 10, it is imperative that all stakeholders openly communicate to find ways to improve current inequalities. The biggest impact can be made by starting small company support programmes, such as restructuring land and township entrepreneur support, and by creating institutions that enable collaboration to tackle community concerns (Makgetla 2020). Iglic et al. (2017) suggest that the use of communication technology improves collaboration among research groups, regardless of geographical barriers, leading to a rise in more successful research projects. All countries, whether developing or developed, are working together to achieve the SDGs through collaboration.

5.5 Conclusion

Research is crucial for addressing social inequalities by providing evidence-based interventions and deepening our understanding of societal issues, particularly complex ones. Through community involvement and policy advocacy, research helps reduce disparities and foster a more equitable and inclusive society. It offers insights into the root causes, recurring patterns, and potential solutions to social inequalities. By applying these findings, stakeholders can collaborate across disciplines to effect positive change. Challenges such as data gaps and policy sustainability remain, but with a focused commitment to evidence-based practices, real equality can be a goal worth striving for. Research is crucial for addressing social inequalities by providing evidence-based interventions and enhancing our understanding of societal issues. Through community involvement and policy advocacy, research helps to reduce disparities and foster a fairer society. It provides insights into the root causes and potential solutions to social inequalities. Research drives social change by offering fresh perspectives that reform policies and empower communities, paving the way toward a society where everyone, regardless of background, has equal opportunities.

References

Akala B (2018) Challenging gender equality in south African transformation policies—a case of the white paper: a programme for the transformation of higher education. S Afr J High Educ 32(3) https://doi.org/10.20853/32-3-1521

Al-Taai S (2021) Sustainable development—a study of its goals, dimensions and methods of achieving. IOP Conf Ser Earth Environ Sci 722(1) https://doi.org/10.1088/1755-1315/722/1/012028

Asian Development Bank (2009) Key indicators for Asia and the Pacific. Asian Development Bank, Mandaluyong

Bebbington J, Unerman J (2015) Achieving the United Nations sustainable development goals: an enabling role for accounting research. Account Audit Account J 31(1):2–24. https://doi.org/10.1108/AAAJ-05-2017-2929

Becker CU (2012) Sustainability ethics and sustainability research. Springer, Dordrecht. https://doi.org/10.1007/978-94-007-2285-9

Biermann F, Kanie N, Kim RE (2017) Global governance by goal-setting: the novel approach of the UN sustainable development goals. Curr Opin Environ Sustain 26–27:26–31. https://doi.org/10.1016/j.cosust.2017.01.010

Chukwu GMO (2015) Participatory evaluation: an action research intervention to improve training effectiveness

Daniels S, Cheng H, Gray C et al (2022) A scoping review of implementation of health-focused interventions in vulnerable populations. Transl Behav Med 12:935–944

de Camara ES, Fernandez I, Castillo-Eguskitza N (2021) A holistic approach to integrate and evaluate sustainable development in higher education: the case study of the University of the Basque Country. Sustainability 13(392):1–19. https://doi.org/10.3390/su13010392

Development Finance International (2023) A call to action to save SDG10: reduce inequalities. Development Finance International, London

Diprete TA, Fox-Williams BN (2021) The relevance of inequality research in sociology for inequality reduction. Socius 7:1–30. https://doi.org/10.1177/23780231211020199

Doyle MW, Stiglitz JE (2014) Eliminating extreme inequality: a sustainable development goal, 2015–2030. Ethics Int Aff 28(1):5–13

Dutonde S (2018) Significance of research in social sciences in India. Int J Curr Eng Sci Res 5(5):71–74

Economic Planning Unit (2017) Malaysia sustainable development goals voluntary National Review 2017, Malaysia

Erismann S, Pesantes MA, Beran D et al (2021) How to bring research evidence into policy? Synthesizing strategies of five research projects in low- and middle-income countries. Health Res Policy Syst 19(29):1–13. https://doi.org/10.1186/s12961-020-00646-1

Ferrell EW (2014) Investigating the effects of facilitating a participatory action research study at a high school with small learning communities. Dissertation,. Loyola University Chicago

Filho WL, Azeiteiro U, Alves F et al (2017) Reinvigorating the sustainable development research agenda: the role of the sustainable development goals (SDG). Int J Sustain Dev World Ecol 25(2):131–142. https://doi.org/10.1080/13504509.2017.1342103

GermanWatch, CAN Tanzania (2017) Collection of SDG factsheets: the 17 sustainable development goals in a nutshell. http://www.ohchr.org/EN/UDHR/Documents/UDHR_Translations/eng.pdf. Accessed 18 Mar 2025

Ghosh E, Nazareth A, Kartha S et al (2023) The missing link between inequality and the environment in SDG 10. In: Bieri S, Bader C (eds) Transitioning to reduced inequalities. MDPI, Basel, pp 19–44

Goodyear VA (2013) Participatory action research: challenging the dominant practice architectures of physical education. University of Bedfordshire

Habib A, Asif M, Haddad H (2024) Does sustainable banking facilitate reducing the SDG-10 in weak rule of law setting? Heliyon 10(2):16. https://doi.org/10.1016/j.heliyon.2024.e24128

Horsley J, Prout S, Tonts M et al (2015) Sustainable livelihoods and indicators for regional development in mining economies. Extr Ind Soc 2(2):368–380. https://doi.org/10.1016/j.exis.2014.12.001

Iglic H, Doreian P, Kronegger L et al (2017) With whom do researchers collaborate and why? Scientometrics 112:153–174. https://doi.org/10.1007/s11192-017-2386-y

International Council for Science (2017) A guide to SDG interactions: from science to implementation

Islam M (2023) The importance of research in the advancement of knowledge and society. Int Res J Basic Clin Stud 8(4):1–3. https://doi.org/10.14303/irjbcs.2023.49

Kim YH, Sting FJ, Loch CH (2014) Top-down, bottom-up, or both? Toward an integrative perspective on operations strategy formation. J Oper Manag 32(7–8):462–474. https://doi.org/10.1016/j.jom.2014.09.005

Kluczyńska U, Kłonkowska A, Bieńkowska M (2024) Researching vulnerable groups: definitions, controversies, dilemmas, and the researcher's personal entanglement. Qual Sociol Rev XX:10–28. https://doi.org/10.18778/1733-8077.20.1.02

Knight SL, Lloyd GM, Arbaugh F et al (2014) Research as a catalyst for change. J Teach Educ 65(2):86–87. https://doi.org/10.1177/0022487113519401

Küfeoğlu S (2022) SDG-10: reduced inequalities. In: Emerging technologies. Springer, Cham, pp 371–384. https://doi.org/10.1007/978-3-031-07127-0_12

Ligozat A-L, Névéol A, Daly B et al (2020) Ten simple rules to make your research more sustainable. PLoS Comput Biol 16(9):1–11. https://doi.org/10.1371/journal.pcbi.1008148

Liu Z, Ye C, Chen R et al (2021) Where are the frontiers of sustainability research? An overview based on web of science database in 2013–2019. Habitat Int 116(August)

Makgetla N (2020) Inequality in South Africa: an overview

Mandela N (2005) Keynote address (live 8 concert in Johannesburg, South Africa) 2 July Available from https://wwwunorg/en/events/mandeladay/legacy shtml#KeyStatements. Accessed 20 January 2024

Maringe F (2014) Theorising research with vulnerable people in higher education: ethical and methodological challenges. S Afr J High Educ 28(2):533–549

Mcmanners PJ (2019) Increasing the impact of sustainability research—a new methodology. J Sustain Res 1:e190008

Michael K, Deshpande T, Ziervogel G (2019) Examining vulnerability in a dynamic urban setting: the case of Bangalore's interstate migrant waste pickers. Clim Dev 11(8):667–678. https://doi.org/10.1080/17565529.2018.1531745

Miya C, Rossier O, Rockwell G (eds) (2021) Right research: modelling sustainable research practices in the Anthropocene. Open Book Publishers, Cambridge

Mosler Vidal E, Laczko F (eds) (2022) Migration and the SDGs: measuring progress—an edited volume. International Organization for Migration, Geneva

NHS Health Scotland (2016) Health inequalities: what are they? How do we reduce them? NHS Health Scotland, Edinburgh

National Institution for Transforming India (2018) Sustainable Development Goals index: Baseline report 2018. https://www.niti.gov.in/sites/default/files/2020-07/SDX_Index_India_Baseline_Report_21-12-2018.pdf. Accessed 18 Mar 2025

Organisation for Economic Co-operation and Development (OECD), International Labour Organisation (ILO) (2019) Tackling vulnerability in the informal economy. OECD Publishing, Paris

Organisation for Economic Co-operation and Development (OECD) (2022) Measuring distance to the SDG targets—Italy. OECD Publishing, Paris, Measuring distance to the SDG targets—Germany. https://doi.org/10.1787/dde28e2c-en

Patel RB, King J, Phelps L et al (2017) What practices are used to identify and prioritize vulnerable populations affected by urban humanitarian emergencies? A systematic review, Oxfam GB, Oxford

Peterson OK (2022) Sustainability and sustainable development research around the world. https://mpra.ub.uni-muenchen.de/115767/. Accessed 21 March 2025

Salas-Zapata W, Ríos-Osorio LA, Cardona-Arias JA (2017) Methodological characteristics of sustainability science: a systematic review. Environ Dev Sustain 52:1127–1140. https://doi.org/10.1007/s10668-016-9801-z

Salvia A, Leal Filho W, Brandli LL et al (2019) Assessing research trends related to the sustainable development goals: local and global issues. J Clean Prod 208:841–849

Shaw R, Sharma A, Takeuchi Y (eds) (2008) Indigenous knowledge and disaster risk reduction: from practice to policy. Nova Science Publishers, New York

South Africa. Department of Environmental Affairs (2016) South Africa environment outlook: a report on the state of the environment. Department of Environmental Affairs, Pretoria, pp 16–26

Stewart F (2013) Approaches towards inequality and inequity: concepts, measures and policies. In: Office of Research Discussion 2013–01, Florence

United Nations (UN) (2015) The millennium development goals report. United Nations, New York

United Nations (UN) (2016) The sustainable development goals report 2016. United Nations, New York

United Nations Educational Scientific and Cultural Organization (UNESCO) (2016) World social science report: challenging inequalities: pathways to a just world. UNESCO, Paris

Vaccarella S, Lortet-Tieulent J, Colombet M et al (eds) (2019) Reducing social inequalities in cancer: evidence and priorities for research. International Agency for Research on Cancer, Lyon

Wilson C, Irvine KN (2013) Bottom-up communication: identifying opportunities and limitations through an exploratory field-based evaluation. Energ Effic 6(1):91–104. https://doi.org/10.1007/s12053-012-9161-y

World Bank Group (2019) Sustainable development goal 10—reduced inequalities: Progress and prospects. World Bank, Washington, DC

World Commission on Environment and Development (1987) Report of the world commission on environment and development: our common future. Oxford University Press, Oxford

Wuelser G, Pohl C (2016) How researchers frame scientific contributions to sustainable development: a typology based on grounded theory. Sustain Sci 11(5):789–800. https://doi.org/10.1007/s11625-016-0363-7

Part II
Social Interventions

6 Rethinking Land Rights to Reduce Inequality: Land Claim of a Tribal Widow From an Indian State

Rai Ganguly

Abstract

Gender inequality in land ownership is a pervasive phenomenon across South Asia. Women face a taboo against working outside the home and a lack of opportunities to create wealth. Inheritance is the primary source of women's property, especially landed property in India. Personal laws recognise women's rights to property without work across all religions, albeit with varying shares. Laws governing land acknowledge both men's and women's equal rights to livelihood. Thus, land as private property and agricultural land are two categories over which women can make claims. However, women from various social groups and religions who work are often excluded from land rights. The case study of a tribal widow's bid for land in the eastern state of West Bengal illustrates how the convergence of two laws may render rural, agrarian women as 'double beneficiaries'. Single women, such as unmarried daughters, or those who are deserted, divorced, or widowed—who are most in need of land—may become landowners through their labour and the inheritance principle, provided that both laws complement each other.

R. Ganguly (✉)
School of Liberal Studies and Humanities, UPES, Dehradun, Uttarakhand, India
e-mail: rai.ganguly@ddn.upes.ac.in

Keywords

Widows · Private property · Land reforms · Legal pluralism · Legal ethnography

6.1 Introduction

This chapter explores how land claims in India are shaped by the intersection of social norms and legal frameworks, both of which are essential for their realisation. India's legal framework is designed to safeguard the unique customs and practices of tribal communities, which often include gender-specific norms governing control over land for males and females. In contrast, statutory land laws aim to be gender-neutral and overlook this differential access to land on the ground. The interplay between land laws and customary practices has significant implications for tribal women's land ownership. This chapter focuses on whether a legally plural system can help reduce the land inequality experienced by women, or whether existing land laws inadvertently reinforce such inequality.

Indeed, women's agency is linked to 'a field of one's own'—an arable plot of land that enhances the status, bargaining power, and well-being of women in South Asia (Agarwal 1994). However, in India, women's subordinate status as landowners indicates that land is a primary determinant of their economic and social inequality. Ownership of land is governed by two distinct

N. V. Roman et al. (eds.), *Social Inequalities Within and Among Countries*, Sustainable Development Goals Series, https://doi.org/10.1007/978-3-032-08741-6_6

categories of laws: those pertaining to the private and public spheres. Family laws govern the private sphere. Essentially, these are codified personal laws derived from religious practices that regulate rights to succession and inheritance of immovable property. In public life, fundamental rights enshrined in the Constitution govern people's right to land. Land is synonymous with livelihood and life, decent work, equal opportunities, and wages. Women's rights over land are adjudged according to one category of law or another.

India is a Hindu majority country. Laws relevant to the private lives of Hindus have undergone multiple amendments to create equality between men and women in marriage, maintenance, adoption, and guardianship. Since 1956, the Hindu Succession Act (HSA) has recognised widows as primary heirs, granting them inheritance rights equal to sons over their deceased husbands' intestate and individual property (Agnes 2011). However, until its amendment in 2005, this central legislation applied only to private property and excluded agricultural land. Land laws—or land reform laws—govern Indian women's rights to agricultural land in the public domain. Under India's federal structure, state legislatures are permitted to interpret the Land Reforms Act of 1955 in ways that reflect regional, religious, and customary practices, as well as geographical variations, land use patterns, and the legal status of each of the 28 states (provinces) and eight union territories (Justice Mullick 2018). The major difference between land as private property and land as an agricultural resource lies in the right to alienate it.

6.2 Legal Interventions and Their Social Relevance: One Right, Two Laws

In this chapter, land reform laws that create land ownership without the right to alienation will be discussed in relation to the Indian state of West Bengal. Its state legislature is known for the systematic enforcement of land reforms, which entitle marginalised groups to land. It also permits the personal laws of individuals to determine the legitimate heir. An ethnographic study of a tribal widow from the marginalised Santhal community exercising her land rights in West Bengal highlights the social control women face in owning and retaining land as a productive resource. Simultaneously, it presents an account of single women from rural and marginalised communities becoming legally active to claim land.

Tribal women's land claims—ranging from kin-group negotiations to courtroom battles—raises a critical question: 'Is there truly a choice to be made between statutory codes and customary practices to secure women's land rights?' Scholars such as Nitya Rao (2007) argue that 'rather than operating in separate fields, the main actor seeks to make interconnections between them. What emerges are *hybrid* systems of jurisprudence', allowing for a degree of manoeuvrability (Rao 2007), emphasis added. Citing Agarwal (1994), Rao further suggests that in patrilineal societies, custom does not necessarily act as a barrier to women's land rights.

Thus, in India, the complexity of laws resulting from affirmative action considers status markers such as age, sex, disability, ethnicity, origin, religion, and economic position to determine land rights. Sustainable Development Goal (SDG) 10, along with its associated targets and indicators—which aim to empower individuals and ensure equal opportunity through the promotion of appropriate legislation (10.3)—must be attentive to existing horizontal stratifications. Recognising these social divisions is essential for formulating effective social protection policies and for progressively achieving greater equality (10.4), rather than overlooking them.

In fact, a study of the enabling and disabling conditions under which women pursue land claims in court can provide valuable insights into how social, economic, and political inclusion of women may be advanced through the hybridisation of legal processes thereby reducing the discriminatory principles embedded in existing practices. In alignment with the ethos of SDG 10, this chapter examines land policy regimes in the eastern Indian state of West Bengal to explore factors that contribute to successful ownership capable of fostering development. It further considers how legal frameworks can be more effectively implemented to promote intersectional equality in land rights, as understood by diverse social groups of women.

6.3 Private Property Versus Tenure Security: A 'Legal Recipe'

Some Indian states have been more successful than others in implementing land reform laws. These reforms emerged as part of broader agrarian reform efforts and were initiated through the abolition of *'zamindari'*—the practice of owning agricultural land as prime private property. Peasants tied to zamindars (landed gentry) cultivated it for generations without land titles. In the 1960s, West Bengal introduced an ingenious 'legal recipe' via the West Bengal Land Reforms Act (WBLRA) of 1955 (Hanstad 2001; Bandopadhyay 2000) to achieve the extraordinary feat of abolishing private property. The state legislature rejected all paperwork connected to land. Instead, government officials accepted verbal testimonies from cultivators to identify households that were the actual tillers of a plot. These individuals were then officially recorded in an indelible 'Record of Rights' as landholders or tenants. Special provisions were included for Scheduled Tribes—historically the primary cultivators of land but excluded from property rights—to ensure that they would not be compelled to transfer their holdings to members of other communities.

The West Bengal legislature recognises both male and female unmarried adults, including divorced or widowed individuals, as tenants entitled to a certain amount of land for subsistence. Furthermore, agricultural land, treated as a special category of property within the public domain, legitimised the sharing of produce in the golden ratio of 3:1. Under this system, known as sharecropping, tenants retain three-fourths of the agricultural yield—provided they bear the costs of seeds, fertilisers, livestock, and farming equipment—while the remaining one-fourth is allocated to the erstwhile landowner. To ensure the continuity of land tenure and prevent land alienation, sharecropped land cannot be legally sold.

In the current context of West Bengal, the tenant or sharecropper is known as *'bargadar'*, while the erstwhile landowner, who receives one part of the produce, is referred to as *'raiyat'*. Agricultural land is considered property under land laws, as it enforces tenure security. Sharecroppers cannot be evicted from the land at the will of the raiyats. The fixed share of the produce is designed to halt exploitation. Moreover, bargadars can also be women, provided they are 'able-bodied' producers. Theoretically, agrarian women's rights to land, irrespective of their religion or ethnicity, are protected under West Bengal land laws. However, their share of land is determined according to personal laws.

The eastern state, known for its high rice production capacity, successfully implemented this legal framework by introducing a stringent land ceiling. A *land ceiling* sets a legal limit on the amount of land a *raiyat's* family of five may retain as private property or for personal cultivation. Any land exceeding the ceiling is vested with the state, with the intent of redistributing the surplus to rightful cultivators in the form of sharecropping land. This policy has created a clear distinction between privately owned land—subject to the provisions of only the HSA involving the right to alienate—and agricultural land, which may be inherited under the same provisions but cannot be legally sold.

West Bengal's flagship land reforms continued into the late 1970s under 'Operation Barga'. The state government vested more land for redistribution by introducing land ceilings on non-agricultural lands, such as factories, fisheries, ponds, and cattle grazing areas. This resulted in multiple lawsuits against the state of West Bengal by aggrieved landowners, who claimed that this violated their right to property. However, the cases were decided in favour of the state, as the right to livelihood—and therefore the right to life—was considered a fundamental right, taking precedence over the right to property (Ganguly 2020).

6.4 Widows' Land Claims: Enabling and Disabling Conditions

In the light of two significant laws enacted in postcolonial, socialist India aimed at promoting agrarian reforms and gender equality, it would appear that Hindu widows could emerge as a distinct and empowered category of landowners. However, research shows exactly the opposite. The relationship between widows and land ownership remains tenuous (Agarwal 1994). Rather than obtaining security, widows are often positioned as the most 'vulnerable category of landowners' or consigned to a state of 'perpetual mourning' (Chen 2000). Scholars have underscored the critical importance of property ownership or access for the overall well-being of rural women (Agarwal 1998). Yet widows—who have historically suffered 'social death' and economic marginalisation following the loss of their husbands—raise a persistent and troubling question: why do they so rarely claim land as a resource, despite legal provisions that ostensibly guarantee their inheritance rights and the promise of a secure livelihood?

Widowhood offers a lens through which to view the contours of land claims, as it is a phase in the life cycle of a woman where she has both social and legal sanctions to property (Rural Development Institute 2009). In West Bengal, only one 'able-bodied' heir can inherit redistributed agricultural land, with no-objection certificates obtained from any other eligible heirs of the original cultivator. Ostensibly, this condition is introduced to prevent *land fragmentation*, a feature that distinguishes agricultural land from private property. Thus, at any given time, only one person from a sharecropping household who is capable of cultivation can have their name registered in the Record of Rights. Therefore, a Hindu widow, whose right to inherit land is recognised in the WBLRA, can claim land as both a legitimate heir and a capable producer. Hypothetically speaking, this rural, agrarian widow emerges as a *'double beneficiary'*.

Moreover, legal sanctions favouring the Hindu widow can be extended to widows from tribal communities as social sanctions. This is because the Indian legal system progressively supports the integrationist perspective, intending to bring socio-culturally isolated tribal communities into the mainstream. The gradual extension of the HSA to such communities is considered legitimate, as statutory law creates greater gender equality in the private domain, where customary norms tend to be most entrenched. Although several Indian states fall under the Fifth Schedule of the Indian Constitution, which protects tribal customs, West Bengal is not among them. Thus, Santhal women, whose labour is integral to agricultural production, are also governed under features of Hindu personal law, traceable in the WBLRA, specifically Section 15(A) that safeguards the rights of cultivation via inheritance upon bargadar's death. Given these clauses, it is pertinent to ask: what are the enabling and disabling conditions under which widows involved in agriculture in West Bengal exercise their land rights by mobilising the legal system?

6.5 Socio-Legal Research and Relevance of the SDGs

Arguably, laws are introduced in India to bring about desirable social change, similar to the SDGs. Existence of legal pluralism—understood as the presence of multiple normative orders governing the same dispute—is recognised by SDG indicator 5.a.2: 'proportion of countries where the legal framework (including customary law) guarantees women's equal rights to land ownership and control'. Just as the implementation of one law affects the outcome of another, the SDGs also recognise that action in one area will influence outcomes in others.[1] Therefore, SDG 5 has implications for SDG 10 regarding the promotion of appropriate legislation and women's agricultural land rights.

Addressing land inequality necessitates focused socio-legal research on the land rights of single women, who are impacted by the intersection of two key legal frameworks. A

[1] www.undp.org

critical first step is to examine the proportion of the 'total agricultural population with ownership or secure rights over agricultural land, by sex'—the first indicator of SDG 5. However, the absence of gender-segregated official data on government-distributed land obscures the extent to which women, as individuals, have actually received land (Ekka 2011:74). Furthermore, in line with the SDG 10.3 target—committed to reducing inequality of outcomes—it becomes imperative to consider SDG 5's second indicator 5.a.1. (b), which assesses the 'share of women among owners or rights-bearers of agricultural land, by type of tenure'. This metric is especially pertinent in evaluating gender disparities in land access and control, also termed tenure security, which is the landholder's perception that rights will be respected by society (Sjaastad & Bromley 2000, cited in Lay et al. 2021).

The following case study adopts a methodology aligned with key aspects of the SDGs, aiming to explore how discriminatory laws governing land ownership might be addressed and how egalitarian legislation can be advanced to strengthen the relationship between agricultural land and tribal women from single-headed households.

6.6 Methodology: Legal Ethnography

The research design does not inquire whether widows go to court; rather, it builds upon the hypothesis that a category of 'litigious widows' exists who claim land as double beneficiaries. Operation Barga introduced the aforementioned legal framework and established quasi-legal bodies known as Block Land & Land Reforms Offices (BL&LROs) across the countryside. Among their many duties, the most important are to update the Record of Rights, oversee all types of land-related disputes concerning redistributed land, prevent illegal buying and selling of land, manage mutations and conversions of land, and conduct on-site verification visits to identify the true tillers of the plot. The land reforms officer and revenue inspectors (RI) also maintain extensive land records, which are a crucial source of data on the number of women *bargadars* and the size of their land parcels in each block. However, this data is not gender segregated. To identify widowed *bargadars* from the digitised land records, the term 'husband' was entered using 'Ctrl+F', resulting in a list of only those women who have inherited land from their husbands. Male *bargadars* and daughters inherit land from their deceased fathers.

The BL&LROs implement the agenda of redistributive justice at the grassroots level, covering all administrative blocks within each district of the state. In addition to examining digitised land records in four blocks of Birbhum district, sharecropping case diaries were also analysed. These diaries contain valuable information on widows who approach the BL&LRO to register themselves as legal heirs. Notably, they also appear as petitioners or as defendant *bargadars,* responding to the summons from their *raiyats.* To gain deeper insight, interviews were conducted with 40 such landowning, litigious widows from four blocks. Each block comprises approximately 200 '*moujas*'—revenue collection units—with each *mouja* encompassing two or three villages. Twelve life histories of Hindu and tribal widows were compiled and corroborated with official accounts from the respective BL&LROs. This ensured a comprehensive legal ethnography covering both the litigants' accounts and case law. The research focused on the period 2010 to 2018, aiming to investigate recent cases of women landowners who assumed responsibility for land cultivation following the deaths of their husbands. A further objective was to assess the impact, if any, of the 2005 amendment to the HSA, which effectively merged agricultural land with private property, on existing land laws. The life history excerpts presented below are particularly noteworthy, as they illustrate the legal and personal aftermath of litigation.

6.7 Single Women as 'Double Beneficiaries': A Case Study

Soni Tudu (name changed) had been widowed for 12 years at the time of theinterview in 2018. Her husband's sister—the only other sibling—also died two years after him, leaving behind four children in her care, the youngest of whom was just a year old. Soni herself had no children. Her sister-in law had married a '*gharjamai*' (husband who moves to his wife's village), who left the village soon after his wife's demise and reportedly died himself in 2016.

During the first three years of her widowhood, Soni cultivated the land with her large family, which included two other adults—the sister-in-law and the *gharjamai* husband. The women had access to five acres of 'barga' land. After that, she employed wage labourers and continued to till the land for another three years. It was at this juncture in 2012, six years after her husband's death, that the *raiyats*[3] started creating legal problems. The land was registered in the name of Soni's father-in-law, and her claim to successorship was challenged by the *raiyat* Mr. Laskar (name changed) on the grounds that registration in the name of the successor should be completed within three months of the recorded *bargadar's* death. Otherwise, the *raiyat* or the RI has the right to appoint an heir. Laskar brought additional charges. In Soni's words:

> After creating trouble, they [the raiyats] went ahead and lodged a case: 'look this woman who cannot cultivate is forcibly occupying the land, and also allowing strangers to cultivate my land'.

This made her appear incapable of cultivating with the help of her own family members, putting her control over the land at stake.

Contrary to this accusation, the sister-in-law's children were old enough and had begun to assist Soni with cultivation. She subsequently lodged a case against her *raiyat* at the BL&LRO in the Bolpur block. As the petitioner, she demonstrated that the RI had listed her son and three daughters as heirs, signifying that they constitute her *family*. Evidently, the inheritance clause was invoked by the widow-heir to counter the accusations made by the *raiyat*. However, several other factors, as gathered from the interview, weigh in.

Firstly, the *raiyat* used both violence and sweet words to deter the widow from cultivating—the only way to prove at the Land Office that the conditions of '*barga*' land ownership have been violated. Soni recalls:

> I was not allowed to step on to my land. We went to the raiyat's house to plough the field, and were told, 'don't come any more, here's one lakh rupees, take it and go'.

Second, Soni not only retained the land despite being offered money, but also went to great lengths to secure her status as a cultivator. The raiyat's case prompted her to undertake the journey to the headquarters in Suri—the District Land and Land Reforms Office.

Third, the sharecropping case file contained no specific information about the land in question. According to Soni, the holding originally comprised 6.5 acres during her mother-in law's time, but had since diminished to 3.5 acres—contrary to the five acres recorded in official documents. This discrepancy reflects a pattern of serious discrimination by the *raiyat* against the *bargadar* widow. An informal mechanism for land transfer, known as *mutual agreement*, exists outside the purview of the state. Under this arrangement, the *raiyat* may sell his land in consultation with the *bargadar*, but is obligated to transfer one-fourth of the sale proceeds to the *bargadar*. This reverses the conventional sharecropping arrangement, in which the *bargadar* delivers one-fourth of the agricultural produce to the *raiyat*. While the share may be.

[3] *Raiyats* were the original owners of the land that was later vested in the state and redistributed. Formerly functioning as intermediaries between landless peasant cultivators and the state—primarily responsible for revenue collection—they now receive a share of the agricultural produce from *bargadars,* the cultivators who hold land titles.

paid in cash, *bargadars* often prefer to receive land in lieu of payment, as it frees them from the ongoing obligation of surrendering a portion of their produce. This informal process, locally referred to as '*barga chharano*'—literally, the relinquishing of sharecropping—is not legally recognised, and as such, the revised landholding is not reflected in official land records.

Before the relationship between Soni and the *raiyats* soured, they asked her to sign off one acre of land in exchange for one lakh rupees. They took her to sign the deed in a car, along with another *bargadar* of theirs, who was also selling his one acre (a *raiyat* may have more than one *bargadar,* depending on the number of families engaged in tilling their land before the application of ceiling). However, the next day, the *raiyat* flatly refused to pay Soni. Going to the local political party did not help, as members had been bribed beforehand—Soni suspects with the money owed to her. However, she discovered that the other *bargadar* had received the due amount from the informal sale. When she demanded to know why she was left out, she was told that the other *bargadar's* land is in their father's name, who is still alive, and therefore their claim to the land is valid, whereas the land she tills is in her deceased father-in-law's name, so she has no claim to it.

Undoubtedly, these three incidents provide insight into why Soni ultimately escalated the matter to the BL&LRO to assert her claim as the rightful heir. The significance of the situation can be seen in the rare instance of a tribal woman bringing a lawsuit against an upper caste *raiyat.* Soni did not activate the system of her own accord; however, when she eventually approached it as an individual, she challenged the system on the basis of her individual capacity for labour—an assertion of her rights that went beyond familial identity or self-defence. This also demonstrates her understanding of the legal rules, which she could use to her advantage by constructing a suitable narrative.

6.8 Social Intervention: Land Rights in New Light

Widows pursuing legal action over property typically presupposes a rupture in kinship ties or a dispute within the family. However, the case study discussed above reveals that women cultivators from female-headed households often fall outside the normative typology of the ideal Hindu family. Furthermore, women who possess agricultural land do not want the same rights as men, as evinced by the actions of the other male *bargadar,* who readily accepted monetary compensation, and the *raiyat,* who engaged in the illicit buying and selling of land rather than remaining content with his share of the agricultural produce. Compensation received from distress sales seldom reflects the actual market value, as the sale of redistributed land is illegal and informal. These transactions largely fall within the domain of propertied classes, who evaluate land based on its potential for alienation, sale, or mortgage—as an income-generating asset—thus subjecting it to the logic of capital. Despite safeguards prohibiting the sale of land, particularly by tribal communities to individuals outside their group, the reduced amount of land in Soni Tudu's possession illustrates that agricultural land continues to exhibit characteristics of private property.

On the other hand, the legal framework governing women's property rights is predicated on the assumption that they seek to *sell* immovable assets, such as land, in times of distress. The HSA acknowledges this possibility by granting women absolute rights over land, thereby formalising their entitlement to alienate it. However, as Soni Tudu's case shows, women's motivations for claiming ownership of agricultural land often diverge from this assumption. Rather than seeking to alienate fertile land, women like Soni seek to belong to it. For such tribal widows, ownership is not solely a legal designation but emerges through control, access, and labour. Soni's litigious engagement demonstrates that she invokes both her right to inherit and her right to cultivate, thereby reinforcing her claim to continued ownership. In this context, ensuring gender-equitable access to land requires a shift in emphasis: social protection policies should prioritise the strengthening of laws that safeguard women's ability to retain land, rather than merely securing their right to alienate it.

It is indeed a paradox to suggest that in order to reduce land inequality and promote land rights among women, laws need to prevent the alienation of land. The legal ethnography discussed here explores a hypothetically clement situation for single women's land ownership, where women are acting on their need—and not entitlement—to legally fight for land. Legal intervention, therefore, precedes any social intervention that may be proposed. There are gaps in socio-legal research that explore best practices among communities across India regarding the gendered devolution of land. Furthermore, suitable literature does not exist on the assessment of land laws from other states of India, which define the sharecropping tenant quite narrowly. Another crucial aspect that needs examination is the effect of the 2005 amendment to the HSA, which reincluded agricultural land within its purview, rendering agricultural land across India as private property. Will this move challenge the land ceiling over land or the autonomy of state legislatures to modify rules according to the tenets of welfarism?

The fact that land is not a homogeneous category, nor are women, calls for an appropriate social intervention to ensure gender equitability in landownership. It may be proposed that each state marks out a certain category of land for female-headed households. Women inhabiting unconventional family structures may initiate multiple uses of land rather than viewing it merely in terms of its market potential. This would move away from the singular logic of patrilineal devolution of wealth and also create room for women to navigate the slippery slopes of customary and statutory laws in India.

References

Agarwal B (1994) A field of one's own: gender and land rights in South Asia. Cambridge University Press, Cambridge

Agarwal B (1998) Widows versus daughters or widows as daughters? Property, land, and economic security in rural India. Mod Asian Stud 32(1):1–48

Agnes F (2011) Family law volume 1: family Laws and Constitutional claims. Oxford University Press, New Delhi

Bandyopadhyay D (2000) Land reform in West Bengal: remembering hare Krishna Konar and Benoy Chaudhury. Econ Polit Wkly 35(21/22):1795–1797

Chen M (2000) Perpetual mourning: widowhood in rural India. Oxford University Press, New Delhi

Ekka A (2011) A status of Adivasis/indigenous peoples land series—4: Jharkhand. Aakar Books, Delhi

Ganguly R (2020) The litigious widow in South Asia: a study in paradoxes. South Asian Surv 27(2):158–171. doi:10.1177/0971523120954863\

Hanstad T, Brown J (2001) Land reform law and implementation in West Bengal: lessons and recommendations. Rural Dev Inst Rep (112) Available at: https://ssrn.com/abstract=826645

Lay J, Anseeuw W, Eckert S, Flachsbarth I, Kubitza C, Nolte K & Giger M (2021) Taking stock of the global land rush: Few development benefits, many human and environmental risks. Analytical Report III. Bern, Montpellier, Hamburg, Pretoria: Centre for Development and Environment, University of Bern; Centre de coopération internationale en recherche agronomique pour le développement; Ger Inst Glob Area Stud; University of Pretoria; Bern Open Publishing. Available at: https://landmatrix.org/resources/land-matrix-analytical-report-iii-taking-stock-of-the-global-land-rush/

Mullick Justice M R (2018) West Bengal land reforms act, 1955. Tenth edition. Kamal Law House, Kolkata

Rao N (2007) Custom and the courts: ensuring women's rights to land, Jharkhand, India. Dev Change 38(2):299–319

Rural Development Institute (RDI) (2009) Women's inheritance rights to land and property in South Asia: a study of Afghanistan, Bangladesh, India, Nepal, Pakistan, and Sri Lanka. RDI, Seattle

Sjaastad E, Bromley DW (2000) The prejudices of property rights: on individualism, specificity, and security in property regimes. Dev Policy Rev 18(4):365–389. https://doi.org/10.1111/1467-7679.00117

Insuring Injustice: Unveiling Inequality in Indian Health Insurance Policies for Women

7

Tanya Goyal

Abstract

Against the backdrop of global commitments to the Sustainable Development Goals (SDGs), particularly universal health coverage [SDG 3.8] to achieve greater equality [SDG 10.3 and 10.4], this chapter explores the state of Indian health insurance policies concerning their coverage of female-specific healthcare needs. This is achieved through a comprehensive document analysis of publicly accessible policy documents from both public and private health insurers. Findings reveal significant gaps in risk and cost coverage across three essential female-specific healthcare concerns: maternity care, gynaecological health, and mental well-being, highlighting discriminatory policies within the Indian healthcare system. This underscores the urgent need for reform in health insurance policies to ensure equitable and justiciable access to healthcare for women. Social interventions, including fostering law-abiding collaborations between government and private sectors to develop effective coverage plans tailored to the specific health needs and social conditions of women, along with greater investment in the healthcare system to enhance service delivery and reduce out-of-pocket expenses (OOPE), become necessary. Ultimately, the goal of achieving greater equality (SDG 10.4) necessitates concerted efforts to eliminate such social inequality, ensuring universal access to quality healthcare, leaving no one behind.

Keywords

Social protection · Health insurance · Universal health coverage · Gender inequalities · Female healthcare · India · Sustainable development

T. Goyal (✉)
Centre for the Study of Social Systems,
School of Social Sciences,
Jawaharlal Nehru University,
Delhi, India
e-mail: tgoyal7820@gmail.com

N. V. Roman et al. (eds.), *Social Inequalities Within and Among Countries*, Sustainable Development Goals Series, https://doi.org/10.1007/978-3-032-08741-6_7

7.1 Introduction

In the Agenda 2030[1] for Sustainable Development, which outlines 17 core goals (United Nations Department of Economic and Social Affairs [UNDESA] n.d.), UN member states—including India—committed to the implementation of national, comprehensive social protection systems (SDG 10.4). This commitment includes the provision of basic services (SDG 1.4[2]), especially universal health coverage (SDG 3.8[3]), to strive for greater equality (SDG 10.3[4] and 10.4[5]). Special attention is di rected toward extending coverage to poor and vulnerable populations (SDG 1.4), especially toward the pregnant women (SDG 1.3.1[6]), and recognition of domestic work (SDG 5.4)[7] (UN 2015).

This way, SDG Target 10.4 highlights the need for social protection systems to attain universal health coverage (SDG 3.8). It is imperative for uplifting vulnerable groups, such as women and the poor, and aims to reduce social inequalities and discrimination in the healthcare system, fostering greater equality within and among countries (SDG 10[8]).

This chapter advances this discussion by examining the Indian healthcare system within the broader context of 'welfare regimes' and health insurance as a crucial component of its social protection system (SDG 10.4). It explores Indian health insurance policies regarding their coverage of women-specific healthcare issues in three critical areas: maternity healthcare, gynaecological illnesses, and mental health issues, through document analysis. Thereby, it discusses potential discriminatory policies, laws, and practices that impact women's access to healthcare and the attainability of SDG 10.3 and 10.4 in this context. Furthermore, the chapter seeks possible social interventions to create a more equitable and inclusive Indian healthcare system, thereby reducing inequalities within and among countries (SDG 10).

7.2 India as a Welfare State

According to Gøsta Esping-Andersen (1990), contemporary welfare states are classified into three types—social-democratic, conservative, and liberal regimes—based on three primary dimensions: state and market relations, social stratification, and social rights.

In line with this classification, India has been characterised as a *social-democratic regime* since its independence from colonialism (1947), as it prioritises social equality and redistributes resources through extensive social programmes funded by progressive taxation (Ellison and Haux 2020). The Indian Constitution promises equality of opportunity to its diverse citizens and prohibits discrimination based on stratifying factors specific to Indian society, mainly gender, class, and caste. This further reinforces the commitment to social

[1] Agenda 2030 is a global framework adopted by the United Nations (UN) and its member nation-states in 2015, aiming to achieve sustainable development for all by the year 2030 through the implementation of 17 goals, known as *Sustainable Development Goals.* (UN 2015).

[2] SDG Target 1.4: By 2030, ensure that all men and women, particularly poor and vulnerable, have equal rights to economic resources and access to basic services. (UN 2015).

[3] SDG Target 3.8: Achieve universal health coverage, including financial risk protection, access to quality essential health-care services and access to safe, effective, quality, and affordable essential medicines and vaccines for all. (UN 2015).

[4] (SDG Target 10.3: Ensure equal opportunity and reduce inequalities of outcome, including by eliminating discriminatory laws, policies, and practices and promoting appropriate legislation, policies, and action in this regard (UN 2015

[5] SDG Target 10.4: Adopt policies, especially fiscal, wage and social protection policies, and progressively achieve greater equality (UN 2015).

[6] SDG Target Indicator 1.3.1: Proportion of population covered by social protection systems, by sex, distinguishing [...] pregnant women, newborns, and the poor and the vulnerable. (UN 2015).

[7] SDG Target 5.4: Recognise and value unpaid care and domestic work through the provision of public services, and social protection policies and the promotion of shared responsibility within the household and the family as nationally appropriate (UN 2015).

[8] SDG 10: Reduce inequality within and among countries. (UN 2015).

justice and egalitarian principles (Government of India 1950).

Later, with its liberalisation, privatisation, and globalisation (LPG) reforms from the 1990s, India embarked on a more *liberal regime*. It gradually shifted towards market-oriented reforms in social sectors, aiming to improve access to quality social services by encouraging private sector participation and promoting market competition (Ravan 2014), while remaining partly oriented towards the social welfare of the vulnerable and poor through a social protection system with *Public-Private Partnership* (PPP) (Christian 2017).

7.3 Social Protection—Indian Health Insurance System

Mandated by SDG 10, Target 10.4, which aims to reduce inequalities, *social protection* refers to a set of programmes and policies designed to eliminate poverty and vulnerability, mitigate exposure to risk, and enhance individuals' capacity to protect themselves against various hazards, disruptions, or income loss (Asian Development Bank 2003). It includes *social insurance*, which safeguards citizens from financial risks and can be provided directly by the state or through subsidised private insurance policies (Morduch 2017). *Health insurance* is one such social insurance, that covers all or a portion of the risk of an individual incurring medical expenses. Like any other insurance models, risk is distributed among multiple stakeholders. By analysing the total health risk and health system expenses across the risk pool, an insurer can establish a regular financing structure, such as a payroll tax or monthly premium, to provide the funds required to pay for the healthcare benefits specified in the insurance agreement (Pekerti et al. 2017).

The Indian health insurance system in particular, is a complex ecosystem comprising government-sponsored (public) health insurance policies and schemes (through PPP) and the private health insurance sector. Government-sponsored health insurance policies and schemes (at the national, state, or block level) provide subsidised to free-of-cost insurance facilities for the poorest and most vulnerable sections of Indian society. In contrast, the private health insurance sector profits significantly from offering various premium health insurance policies that cater to the diverse needs and preferences of the Indian middle class, including lower, middle, and upper strata (Christian 2017). The system is designed as a three-tier structure consisting of primary (basic general emergency services), secondary (specialised (care focused on specific needs/medical field) services), and tertiary (super-specialised (focused care within a subspecialty) services) facilities. The main approach is to provide subsidised primary care through government-sponsored schemes funded by general tax revenue, supplemented by the strategic purchase of secondary and tertiary care services from both public and private company insurers (Ministry of Health and Family Welfare 2017).

However, existing literature indicates that the Constitution of India and its laws do not explicitly recognise health and healthcare as rights. Instead, it imposes a directive obligation on the government to ensure quality healthcare for all without any discrimination through PPP (Gangolli et al. 2005:281), which is not justiciable or enforceable in case of violations (Chatterjee and Sheoran 2007: 28). Healthcare insurers then, have the autonomy to decide whether to insure something or not and at what rate, based on their profit, demand, and supply over equitable access (Forum for Medical Ethics Society 2015).

In addition, the literature highlights that currently, only 2.26% [9]of the total union budget expenditure in India is allocated to the health sector (Rathore 2023). Furthermore, government-sponsored programmes constitute only a small proportion—just over 1%—of the total allocated budget, which has proven inadequate to meet the specific needs of the vast vulnerable, lower-income population (Selvaraj et al. 2022: 79). This pushes them to seek OOPE when having

[9]The figures and data presented in this chapter correspond to official data available for India up to 2023, consistent with the study timeframe.

no option than using unregulated private healthcare services due to shortage and inefficiency of public-healthcare facilities (Sriram and Khan 2020: 19).

Females, as one such vulnerable, low-income group with low literacy rates, constitute approximately half of the overall population in India (National Statistical Office 2022). They contribute significantly to the economy through their 'unpaid labour' (domestic work) and 'reproductive work' (care work) (Lingam and Yelamanchili 2011). However, they remain disproportionately affected by these systemic gaps, facing low accessibility to efficient and costly facilities needed for the treatment of female-specific illnesses, including emergencies such as essential maternity treatment and care (Selvaraj et al. 2022).

To instantiate, various studies have noted that government-sponsored programme schemes providing subsidised general health insurance to improve health outcomes for vulnerable populations (Kumari and Manisha 2023) become inadequate due to their eligibility criteria. This includes a low female acceptance rate, stemming from costly treatment and procedures. Additionally, these schemes offer only a basic insurance plan on a *family basis* for general illness coverage, which grants authority to the head of the family or household regarding which illnesses to insure and whom to register under the plan (Ram Prakash and Lingam 2021). With male-biased decisions and the allocation of household resources in India (Desai 2009), women are sidelined from accessing basic health insurance services, thereby exacerbating *gender inequality* and its *intersection* with accessing healthcare services and claiming insurance. This situation also *leads to discrimination* against women as a vulnerable group with their own unique healthcare needs (World Health Organisation [WHO] 2021). Hence, instead of ensuring women have access to female-specific quality healthcare, Indian public health insurance schemes simply end up financing their own gaps, failing to address vulnerability and these *social inequalities* that are their primary objective (Desai 2009).

The country consistently ranks among the five lowest-performing nations globally for women's health and survival (World Economic Forum [WEF] 2023). Even when Indian public-healthcare programmes and schemes are observed to focus on women's healthcare needs solely in terms of reproduction (Forum for Medical Ethics Society 2015), the infant and maternal mortality rates remain higher, compared to other countries in the Asian region that have similar income levels, and rates of economic growth (WHO 2023).

Even though numerous analyses exist of such family-oriented public-healthcare programmes and policies implemented through PPP, which cover only specific income, occupation, or regional groups (Central Bureau of Health Intelligence 2022), the existing literature falls short in providing critical analysis of the broader Indian health insurance system. As a result, scholarly understanding remains partial and fragmented. Overlooked in many studies is the complex landscape comprising a wide range of Indian public and private insurance companies that are Present Across Nation (PAN), along with their per-person health coverage policy plans and documents. The lack of attention to these elements is particularly troubling given women's distinct healthcare needs and their existing situation of overall access to comprehensive health services. To address this gap, the present study comprehensively and critically examines these existing insurance policies with the aim of evaluating how effectively the current Indian health insurance system responds to the unique needs of women as a vulnerable group. Such an analysis is essential for identifying potentially discriminatory policies and practices that perpetuate healthcare inequities and discrimination, while also providing an evidence base for related social interventions aimed at creating a more equitable healthcare system.

7.4 Methodology

Employing document analysis,[10] this study reviews publicly accessible policy documents of all four public sector general and life insurance company (s), and all seven private standalone health insurance companies (including all three female-specific health insurance policies) available under the category of 'comprehensive health insurance providers', that are available PAN India, and are registered and approved by the Insurance Regulatory and Development Authority of India (IRDAI) as of 2023 (Kumar et al. 2022).

The data analysed spans the period from 2021 to 2023. The list of healthcare insurers was obtained from the IRDAI's (2023) electronic database. The wording of the published policy documents was retrieved from the publicly accessible official websites of each insurer. For each company, three categories of policy plans were combined for analysis: the basic plan (the cheapest), the comprehensive plan (the most expensive) with optional add-on premium covers, and the female-specific plan. This categorisation allows for a detailed and comprehensive evaluation of the scope of healthcare coverage and associated costs. To preserve confidentiality, the identities of the insurers have been anonymised in the study.

The unit of analysis is the coverage of healthcare issues specific to females in these policies, classified under three themes: (1) maternity (pre- and post-hospitalisation) and newborn healthcare, (2) gynaecological diseases and their diagnostic tests, and (3) mental health issues. These three categories were chosen prior to data collection, as they are found to be the most common health concerns specific to females (WHO 2021).

7.5 Findings and Discussion

The main finding reveals a significant lack of coverage for female-specific healthcare needs across both public and private health insurers. While a few private insurers acknowledge these needs by offering female-specific health insurance plans, such efforts remain limited. Further details are provided below.

7.5.1 Public Sector General and Life Insurer(s)

(1) Maternity and Newborn Healthcare: Strikingly, none of the public sector general or life insurers extend comprehensive coverage to maternity-related diagnostics and routine check-ups. Where coverage for pregnancy-related services—such as normal delivery, caesarean section, abortion, and ectopic pregnancies—is provided, it is typically limited to 10% of total expenses. One insurer offers 10% coverage without a premium, but such instances are exceptions. This limited coverage too is available only under selective policies, often subject to lengthy waiting periods (ranging from one to three years) and the purchase of additional premium packages. Newborn healthcare and vaccinations follow a similar pattern, with coverage restricted to approximately 5% of total costs. In cases where the required medical treatments are unavailable in empanelled network hospitals, *insurers may recommend that patient(s) seek private healthcare services,* often at their own expense.

(2) Gynaecological Diseases and Their Diagnostic Tests: Coverage for gynaecological conditions is minimal and inconsistent. All insurers uniformly exclude coverage for sexually transmitted diseases (STDs). Among the reviewed insurers, three provide coverage for polycystic ovary syndrome (PCOS) and polycystic ovarian disease (PCOD), but only after a two-year waiting period, and only one insurer covers a range of gynaecological issues after a two-year waiting period, depending on selective premium covers. Only one insurer offers broader coverage for gynaecological diagnostic tests, though this is limited to once every two years and explicitly excludes reproductive health tests.

(3) Mental Health Issues: Mental health coverage is highly restricted among general insurers. Most insurers only provide coverage

[10] The "READ" approach was followed in this document analysis (Dalglish et al. 2020).

in cases where the mental health condition poses a threat to life and requires hospitalisation. There are some variations in coverage among insurers, with one excluding mental health coverage entirely. Public sector life insurers—despite their widespread preference among Indian families and households (The Indian Express 2021)—do not offer any significant differentiation in mental health coverage.

7.5.2 Private Sector Standalone Health Insurers

(1) Maternity and Newborn Healthcare: Compared to public insurers, the majority of standalone health insurers provide coverage for newborn vaccinations with a standard premium (five out of seven). However, maternity healthcare is excluded entirely by two insurers unless specific premium plans are chosen. Even then, coverage is subject to conditions such as waiting periods ranging from 1 to 4 years and limits after two pregnancies.

(2) Gynaecological Diseases and Their Diagnostic Tests: There is no coverage for STDs; if any exists, it is not without a premium. Similarly, three insurers exclude coverage for PCOS/PCOD, other gynaecological diseases, and STDs, while others include it only if it becomes life-threatening or if a specific premium plan is taken, which comes with a waiting period of 2 to 4 years. Three out of seven insurers include diagnostic tests only with premium coverage.

(3) Mental Health Issues: Four insurers offer coverage for mental health issues only after hospitalisation and with a premium (one offers this coverage only with a lifetime premium). The other three insurers exclude coverage for mental health entirely.

7.5.2.1 Female-Specific Health Insurance Offered by Private Sector Standalone Health Insurers

Female-specific health insurance plans appear to provide better coverage for female-specific needs, although they tend to be costlier.

(1) Maternity and Newborn Healthcare: There is not much difference in cost coverage here, although one insurer also covers pre- and post-hospitalisation consultations up to 10%. All plans cover pregnancy delivery, including normal deliveries and certain complications, subject to the premium plan and a waiting period of 1–3 years for up to two pregnancies, while one insurer covers complications up to 25% of the total cost (not exceeding 5 lakhs INR). Another insurer includes premium coverage for surrogate mothers, oocyte donors, and even legally adopted children.

(2) Gynaecological Diseases and Their Diagnostic Tests: STDs are excluded in all policies. PCOS/PCOD and other gynaecological diseases are partially covered under premium plans, with waiting periods of 2–4 years. One insurer covers these conditions only at a life-threatening stage, providing 25% coverage. Gynaecological diagnostics are covered by two insurers, one without a premium and another with an annual premium and a 30-day waiting period.

(3) Mental Health Issues: Coverage is limited to life-threatening conditions that require hospitalisation for most insurers. There are variations in coverage, with one insurer opting not to cover mental health issues at all.

The above data indicate that the majority of both private and public health insurance policies fail to adequately cover treatment costs for, or even include, female-specific illnesses. Even when coverage is available, it typically accounts for only about 10% of total costs, and is frequently subjected to monthly premiums, additional costs, and extended waiting periods—even for basic maternity care add-ons. Premium plans offer only marginal improvements, such as slightly reduced waiting periods and modest increases in financial coverage (averaging 5–10%), while adhering mostly to the risk coverage of basic plans.

Similar issues of limited coverage,[11] coupled with severe unavailability of required specialised facilities[12] and healthcare staff, have been noted

---

[11] Includes both in-patient and out-patient care facilities and medication charges.

[12] Cf. National Health Accounts Estimates for India (Ministry of Health and Family Welfare 2023)—Figure 5,

by the National Health Authority, India (2019), in current PAN India government health insurance programme schemes.[13] These schemes are specially designed to provide subsidised general health insurance plans to vulnerable and low-income populations, aiming to achieve the vision of *universal health coverage* as outlined by the Ministry of Health and Family Welfare (2017), to uphold the SDGs' core commitment to *leave no one behind*. However, in practice, these schemes risk inadvertently perpetuating gender discrimination by failing to address the specific healthcare needs of women.

As a result, more than 50% of total health expenditures in India rely on user fees or OOPE—one of the highest and most catastrophic rates[14] globally (World Bank Group 2023). This disproportionately affects millions of poor and vulnerable populations, particularly women with unique and often costly healthcare needs, pushing those in the lowest income brackets further below the poverty line (Selvaraj et al. 2022). Consequently, OOPE has become a significant contributor to poverty (Forum for Medical Ethics Society 2015), further increasing social inequalities.

While intersectionality is not the focus of this chapter, region-specific inequalities are evident. Almost 70% of insurance-affiliated active and efficient hospitals (both public and private) and offices are located in Tier 1 urban cities (developed metropolitan regions with higher population density) of India, which comprise less than a third of the nation's total population (O'Neill 2024). This concentration means that vulnerable groups, particularly women in Tier 2 and Tier 3 cities, as well as rural areas, suffer from delays in registering for and claiming insurance and package benefits (Kumar et al. 2022:55).

Thus, even if a sufficiently large subsidy and the availability of general and one-person-based insurance policies might increase the registration of some women in healthcare insurance plans, benefits are still not accessible to them (Dupas and Jain 2021). This indicates that the current active healthcare insurance policies fail to cover the actual targeted vulnerable groups, creating grounds for discrimination, with no justiciability due to the lack of laws in this regard. Consequently, these programmes then become high-cost, profitable premium policies that are 'affordable only to affluent low-risk individuals—policies affordable only because they exclude lower-income, higher-risk individuals', thereby increasing existing social inequality (Nadash et al. 2017).

Similar patterns can be observed in various other lower-middle and low-income Asia-Pacific countries, where OOPE accounts for nearly half of total health spending, and female-specific treatment facilities are often underfunded or neglected in health insurance policies. Women commonly face challenges accessing healthcare due to financial constraints and the distance to health facilities. The maternal mortality ratio in the region averages almost three times that of upper-middle-income countries and more than 15 times that of high-income Asia-Pacific countries and territories. India and Pakistan alone accounted for 15% of the world's maternal deaths (Organisation for Economic Co-operation and Development [OECD] and WHO 2022). This underscores the global inequality between low- and high-income countries in providing universal healthcare access to such vulnerable populations (UN 2023).

As practiced in many of OECD countries[15] and aligned with the provisions of SDG 3.8, inclusive and comprehensive health insurance benefits should be guaranteed automatically as a legal entitlement for all individuals, based on their health

which illustrates a declining share of specialised health care in current government health expenditure (in the reference section).

[13] As has been noted earlier in the third section of the chapter.

[14] A household is considered to have experienced catastrophic health expenditure if its OOPE exceeds 10% of the total household income (Selvaraj et al. 2022).

[15] The Organisation for Economic Co-operation and Development (OECD) is a group of 38 countries that aims to promote economic growth, prosperity, and sustainable development among its member countries. These countries are characterised by high-income levels, strong democratic institutions, and a commitment to free market principles. They are spread across different continents, including North America, Europe, Asia, and the Pacific (OECD 2024).

needs, irrespective of socio-economic status, ability to pay, or geographic location. Such an approach is essential to achieving universal health coverage and reducing health-related inequalities, as emphasised in SDG 10.3 (McVay 2023).

7.6 Social Intervention

As we have seen, there are profound barriers to women's access to essential, affordable, and quality healthcare, leading to a significant gap in India's provision of universal healthcare coverage to all its citizens and the achievement of greater social equality. This indicates an urgent need for targeted and actionable social interventions that could counteract the structural dynamics of inequality holding women back, which will be explored in this section.

While the majority of countries have embraced Agenda 2030, the path to achieving most of the SDGs by 2030 presents considerable challenges for many low and lower-middle-income countries, including India, which represents a substantial portion of the global population (UN 2021). For India, major challenges remain unaddressed, especially regarding SDG 3 (good health and well-being) and SDG 5 (gender equality), with slow growth and low commitment creating obstacles to achieving SDG 10 which calls for reduced inequalities, with little data available (Sachs et al. 2023: part 2). The COVID-19 pandemic exacerbated these issues, making India one of the lowest-ranking nations in the global health-security index and in female mortality rates. Currently, the SDGs are not included in the national budget plan, nor are their strategies included in sectoral action plans (Sachs et al. 2023: part 3). Furthermore, due to the absence of a fundamental right or law for health, there are multiple regulatory loopholes, leading to an inefficient healthcare system with inadequate services and insurance plans for its vulnerable population, especially women (Chatterjee and Sheoran 2007).

This presents a hurdle to building trust and cooperation among countries to work together towards sustainable development (Sachs et al. 2023: part 3). Therefore, regular and open monitoring and evaluation of progress—alongside a strong emphasis on pathway analysis in relation to SDG indicators—are essential for governments to design partnership-based, long-term investment strategies that advance national progress toward achieving SDG 10 (Sachs et al. 2023: part 1). National governments should draw on the experiences of other countries, adapting relevant insights to their own healthcare contexts through collaboration with diverse local stakeholders.[16] This should be accompanied by the integration of the SDGs—especially SDG 3 and 5—into national and local budgetary plans, and into the development of strategies that are both politically feasible and operationally viable.

Health and human rights advocacy in India need to intensify efforts to transform the obligatory Directive Principles on health into an autonomous right through rigorous judicial activism. This includes filing Public Interest Litigations and gathering testimonials regarding the denial of the right to health (Chatterjee and Sheoran 2007:28) to support evidence-based standard laws and regulations for healthcare, applicable to both the public and private sector healthcare service providers (Ministry of Health and Family Welfare 2017).

Given that local governments are responsible for implementing universal health systems at the ground level, as mandated by national commitments, there is a need for increased healthcare budget allocations and adjustments to them from the national government (Sachs et al. 2023: part 1). Enhanced funding would incentivise collaboration among local governments, the public health sector, and private companies to improve the distribution of functional healthcare facilities—including preventive in-patient and out-patient care[17]—as well as infrastructure and healthcare personnel. Without investments in quality, nations risk inefficient expansion and the provision of

[16] Includes local government bodies, social workers, and vulnerable communities in target.

[17] Evidence suggests that outpatient costs can be as debilitating as in-patient events' costs. Hence, preventive and diagnostic care should be encouraged and subsidised (Lagomarsino et al. 2012:939).

unnecessary or substandard healthcare services (Lagomarsino et al. 2012:940).

Health insurance policies originating from PPP then, could offer a few but effective coverage plans and facilities based on robust data, minimising administrative costs compared to numerous low coverage plans (Mossialos et al. 2016). Here, national governments need to get involved to create effective insurance systems that pool risks according to the needs of females at the local level, through dedicated resources, female leadership in decision-making, and enhanced capacity for active regulations and routine data collection, especially from the target population (Lagomarsino et al. 2012:940).

Working with local social workers and female researchers can gather such region-specific data and applied social research outputs on women's unique healthcare needs, enabling insurers to tailor policies to fit the actual risk-induced population. Furthermore, insurers (especially public ones) can also use their existing services to disseminate comprehensive information about health policies to raise public awareness and benefit hard-to-reach communities. Non-governmental organisations can also facilitate, monitor, and advocate for the proper functioning of schemes and the availability of benefits to actual vulnerable populations by coming together and collaborating with government programmes.

Lastly, a dedicated global commitment to SDG 10—supported by transparent and shared cooperation and responsibilities among countries—can exert constructive pressure on stakeholders, including national and local governments and civil society, to target the improvement of female healthcare for their social inclusion (SDG 10.2). This entails ensuring equal opportunities for good health by addressing their unique needs, eliminating discriminatory laws, policies, and practices, and promoting equitable legislation and targeted interventions (SDG 10.3). Such efforts are essential to advancing greater equality within and among countries (SDG 10.4), a goal that lies at the heart of Agenda 2030 (UNDESA and World Bank Group 2019) and is central to achieving sustainable development, improving the world for all (United Nations Regional Information Centre for Western Europe 2024).

7.7 Conclusion

This chapter explored discrimination against women in Indian health insurance policies, relating it to existing laws, policies, and practices. It highlighted the minimal coverage for female-specific healthcare concerns, which contributes to social inequality. Furthermore, it underscored the critical role of social protection (SDG 10.4), comprising an inclusive and efficient healthcare system with effective health insurance services, as a pivotal social intervention to ensure universal health coverage (SDG 3.8) and achieve greater equality (SDG 10.4), as committed to by India.

> Unless we act now, the 2030 Agenda will become an epitaph for a world that might have been.
> António Guterres, Secretary-General of the United Nation (UN 2023).

The concern is not whether the policy is available, but whether it is applied to its fullest potential, serving the needs of a country's healthcare system by measuring what proportion of the population that requires healthcare interventions actually receives its benefits (WHO 2004:530).

References

Andersen GE (1990) The three worlds of welfare capitalism. Princeton University Press, Princeton

Asian Development Bank (2003) Social protection: Asian Development Bank. https://www.adb.org/sites/default/files/institutional-document/32100/social-protection.pdf.

Central Bureau of Health Intelligence (2022) National health profile 2022. In: Ministry of Health and Family Welfare, government of India, 17th edn, New Delhi. https://cbhidghs.mohfw.gov.in/WriteReadData/l892s/94203846761680514146.pdf. Accessed 18 Mar 2025

Chatterjee CB, Sheoran G (2007) Vulnerable groups in India. Centre for Enquiry into Health and Allied Themes, Mumbai

Christian A (2017) The Routledge international handbook to welfare state systems. Taylor & Francis, London, p 26

Dalglish SL, Khalid H, McMahon SA (2020) Document analysis in health policy research: the READ approach. Health Policy Plan 35(10):1424–1431. https://doi.org/10.1093/heapol/czaa064

Desai S (2009) Keeping the "health" in health insurance. Econ Polit Wkly 44(38):18–31

Dupas P, Jain R (2021) Women left behind: gender disparities in utilization of government health insurance in India. Natl Bur Econ Res Work Pap Ser. https://doi.org/10.3386/w28972

Ellison N, Haux T (2020) Welfare states and social policies. In: Ellison N, Haux T (eds) Handbook on society and social policy. Edward Elgar Publishing, Cheltenham, p 227

Forum for Medical Ethics Society (2015) National health policy 2015. Econ Polit Wkly 50(36)

Gangolli LV, Duggal R, Shukla A (2005) Review of healthcare in India. Centre for Enquiry into Health and Allied Themes, Mumbai

Government of India (1950) Constitution of India. Government of India, New Delhi

Insurance Regulatory and Development Authority of India (IRDAI) (2023) Handbook on Indian insurance statistics. IRDAI, Hyderabad

Kumar GRS et al (2022) Annual report. Insurance Regulatory and Development Authority of India, Nanakramguda, Hyderabad

Kumari S, Manisha M (2023) Status of women's health Insurance in India: an appraisal. Econ Polit Wkly 58(51)

Lagomarsino G, Garabrant A, Adyas A et al (2012) Moving towards universal health coverage: health insurance reforms in nine developing countries in Africa and Asia. Lancet 380(9845):933–943. https://doi.org/10.1016/s0140-6736(12)61147-7

Lingam L, Yelamanchili V (2011) Reproductive rights and exclusionary wrongs: maternity benefits. Econ Polit Wkly 46(43)

McVay D (ed) (2023) Germany: Health system overview. https://healthsystemsfacts.org/national-health-systems/bismarck-model/germany/germany-health-system-overview/. Accessed 31 Mar 2024

Ministry of Health and Family Welfare (2017) National health policy. Government of India, New Delhi

Ministry of Health and Family Welfare (2023) National health accounts es*timates* for India (2019–20) released. https://pib.gov.in/PressReleaseIframePage.aspx?PRID=1919582.

Morduch J (2017) Economics and the social meaning of money. In: Bandelj N, Wherry FF, Zelizer VA (eds) Money *talks*: explaining *how money really works*. Princeton University Press, Princeton, pp 25–38

Mossialos E, Wenzl M, Osborn R et al (eds) (2016) 2015 *international profiles* of *health care systems*. The Commonwealth Fund, New York

Nadash P, Doty P, von Schwanenflügel M (2017) The German long-term care insurance program: Evolution and recent developments. Gerontologist 58(3):588–597. https://doi.org/10.1093/geront/gnx018

National Health Authority (2019) Ayushman Bharat-packages and rates: about Pradhan Mantri Jan Arogya Yojana (PM-JAY). Government of India, New Delhi

National Statistical Office (2022) Women and men in India 2022. Ministry of Statistics and Programme Implementation, Government of India, New Delhi

O'Neill A (2024) India: Degree of urbanization from 2012 to 2022. https://www.statista.com/statistics/271312/urbanization-in-india/.

Organisation for Economic Cooperation and Development (OECD) (2024) About OECD. https://www.oecd.org/about/. Accessed 17 June 2024

Organisation for Economic Co-operation and Development (OECD), World Health Organisation (WHO) (2022) Health at a glance: Asia/Pacific 2022: measuring progress towards universal health coverage. OECD Publishing, Paris

Pekerti A, Vuong QH, Ho TM, Vuong TT (2017) Health care payments in Vietnam: patients' quagmire of caring for health versus economic destitution. Int J Environ Res Public Health 14(10):1118. https://doi.org/10.3390/ijerph14101118

RamPrakash R, Lingam L (2021) Why is women's utilization of a publicly funded health insurance low?: A qualitative study in Tamil Nadu. India BMC Public Health 21:350

Rathore M (2023) India: Budget share for health sector 2023. https://www.statista.com/statistics/1356501/india-budget-share-for-health-sector/. Accessed 30 Mar 2024

Ravan SV (2014) Impact of LPG on Indian economy. Prime Int Res J 1(4):21–33

Sachs JD, Lafortune G, Fuller G, Drumm E (2023) Implementing the SDG *stimulus*. Sustainable Development Report 2023. Dublin University Press, Dublin

Selvaraj S, Karan KA, Srivastava S, Bhan N, Mukhopadhyay I (2022) India health system review. World Health Organisation Regional Office for South-East Asia, New Delhi

Sriram S, Khan MM (2020) Effect of health insurance program for the poor on out-of-pocket inpatient care cost in India: Evidence from a nationally representative cross-sectional survey. BMC Health Serv Res 20(1):839. https://doi.org/10.1186/s12913-020-05692-7

The Indian Express (2021) LIC marks 66 years of Existence: 'True to nationalisation objective, taking insurance to all at reasonable cost'. https://indianexpress.com/article/business/companies/life-insurance-corporation-of-india-lic-66-year-existence-7482928/.

United Nations (UN) (2015) Transforming our world: the 2030 agenda for sustainable development. UN, New York

United Nations (UN) (2023) The sustainable development goals report 2023: special edition-towards a rescue plan for people and planet. UN, New York

United Nations Department of Economic and Social Affairs (UNDESA) (n.d.) The 17 goals|sustainable

development. United Nations. https://sdgs.un.org/goals. Accessed 26 Mar 2024
United Nations Department of Economic and Social Affairs (UNDESA), World Bank Group (2019) Sustainable development goal 10—reduced inequalities: Progress and prospects. Sustain Dev. https://sustainabledevelopment.un.org/content/documents/21453SDG_10_EGM_2019_concept_note_30Jan_consolidated.pdf. Accessed 17 June 2024
United Nations Regional Information Centre for Western Europe (UNRIC) (2024) Goal 10: reduce inequality within and among countries. UN, New York
World Bank Group (2023) Out-Of-Pocket Expenditure (% Of Current Health Expenditure)-India. https://data.worldbank.org/indicator/SH.XPD.OOPC.CH.ZS?end=2020&locations=IN&name_desc=false&start=2020&type=shaded&view=map.
World Economic Forum (WEF) (2023) Global gender gap report 2023. WEF, Geneva
World Health Organisation (WHO) (2004) Regional overview of social health Insurance in South-East Asia. WHO, Geneva
World Health Organisation (WHO) (2021) Gender and health. WHO, Geneva
World Health Organisation (WHO) (2023) Indicators. WHO, Geneva

8 Losing Home, Occupation, and Language: A Case Study of Afghan Evacuees in Germany After 2021

Sajia Behgam Amin and Dagmar Oberlies

Abstract

In response to the Taliban's takeover of Afghanistan in 2021, Germany, alongside several other countries, pledged to welcome Afghan refugees, particularly human rights defenders and their families. According to the United Nations High Commissioner for Refugees, Afghans represent the third-largest refugee group globally, following Syrians and Ukrainians. Of the 5.7 million Afghan refugees worldwide, nearly 400,000 reside in Germany, approximately one-third of whom are women. Facilitating safe and responsible migration through policies is one of the key targets of the Sustainable Development Goals (SDGs), namely SDG 10.7, which is aimed at reducing global inequalities. Focusing on the aftermath of the 2021 evacuation, this study explores the situation of Afghan evacuees and the extent to which German policies and regulations have responded to the needs of the Afghan diaspora.

Keywords

Afghan refugees · Migration policies · Integration challenges · Family reunification · Sustainable development goals (SDG 10.7)

> We lost our home, which means the familiarity of daily life. We lost our occupation, which means the confidence that we are of some use in this world. We lost our language, which means the naturalness of reactions, the simplicity of gestures, the unaffected expression of feelings.
>
> — Hannah Arendt (1943: n.p.)

8.1 Introduction

In May 2021, the Taliban launched a major offensive in Afghanistan following the withdrawal of the first United States (US) troops from the country. In the subsequent months, they captured one province after another, culminating in the fall of Kabul on 15 August 2021. In the immediate aftermath, more than 120,000 frightened people were airlifted from Kabul International Airport.

As of April 2024, more than 20,000 support staff who had worked with the German military, along with an additional 12,900 'defenders of a democratic Afghanistan', were evacuated to Germany with their immediate families. Nearly 50,000 Afghans are currently in the process of applying for individual asylum. Moreover, 2000

S. B. Amin (✉) · D. Oberlies
Faculty of Health and Social Sciences, Frankfurt University of Applied Sciences, Frankfurt, Germany
e-mail: sajia.behgam-amin@fb4.fra-uas.de

D. Oberlies
e-mail: oberlies@fb4.fra-uas.de

N. V. Roman et al. (eds.), *Social Inequalities Within and Among Countries*, Sustainable Development Goals Series, https://doi.org/10.1007/978-3-032-08741-6_8

dependent family members have been admitted, and no deportations to Afghanistan have taken place since 2021. At present, approximately 400,000 Afghans live permanently in Germany, making up 3% of the country's foreign national population.

This case study aims to deepen understanding of how migration policies can contribute to reducing inequalities, as outlined in Sustainable Development Goal (SDG) 10, and to evaluate the effectiveness of migration management from the perspective of its beneficiaries. It focuses on the experiences of Afghans evacuated to Germany after 2021, juxtaposing their personal narratives with the provisions of German migration law and policies.

8.2 Literature Review

The Federal Office of Migration and Refugees (BAMF) and the Institute for Employment Research (IAB) conduct regular studies on the expectations and experiences of refugees regarding their integration into the labour market, education system, and German society. In an initial survey, Brücker et al. (2016) found diverse educational backgrounds within the refugee population and a strong motivation to integrate into the German education system and labour market. Language acquisition was crucial and depended on age, level of education, length of stay, intention to remain, contact with locals, and media consumption (BAMF 2021b).

The particular situation of refugee women was further examined (Böhm 2019; BAMF 2021a; DIW Berlin 2023): Their participation in the labour market is lower due to their educational backgrounds, family structures, and caregiving responsibilities. Despite these challenges, four out of five refugee women attended language courses and achieved at least an intermediate level of proficiency. Traumatic experiences, depression, and a lack of childcare options affected language learning. Special programmes and language courses, along with improved German skills, have significantly increased women's participation in the labour market. Compared to male refugees, women have fewer contacts with Germans and participate less in education and employment. Their leisure activities are also limited, and contacts often arise through their children.

Studies on Afghan refugees are limited: In a study involving 32 Afghans evacuated to the USA, Nylen et al. (2023) explored the experiences and perceptions of Afghan evacuees, as well as their immediate and long-term needs. They concluded with recommendations to adopt a non-Western approach to the concept of 'family', empower displaced populations, improve training and expectation management, coordinate with non-governmental organisations (NGOs), incorporate gender considerations into all efforts, establish an easily accessible information portal, organise refugee-centred job fairs, and integrate culturally sensitive mental health care into refugee health services. Safi looked at Afghans in Germany and the impact their migration had on their culture (Safi et al. 2023), as well as their engagement in diaspora organisations (Safi and Czaika 2024). Several studies have examined health-related issues affecting Afghan refugees. A literature review on the health situation in Europe identified not only infections, anaemia, diabetes, and obesity, but also mental health challenges, including culture shock, feelings of displacement, low socioeconomic status, and limited access to healthcare facilities (Matsangos et al. 2022). Economic data from the USA indicate that Afghan refugees face considerable hardship, recording the lowest income levels among seven comparison groups (Stempel and Alemi 2021). In Australia, a study on language acquisition among female Afghan refugees highlighted major barriers related to self-esteem and motivation (Sharifian et al. 2020).

8.3 Methodology

The study employed a qualitative methodology with an exploratory design (Swedberg 2020).

Focus group discussions (FGDs) were conducted with a group of 34 Afghan evacuees (Currle 2011; Kolendowicz 2017). The invitation to participate was distributed through various Afghan networks on social media, specifically among participants in a further training programme at Frankfurt University of Applied Sciences (UAS) (called STEPS-A). Initially, 40 people agreed to participate, but only 34 attended. Discussions were conducted online because some participants could not travel or leave their children behind. Technology was not an issue, but interaction may have suffered; it was lower in the men's group, possibly because the facilitator was a woman.

The four focus groups consisted of human rights activists, former civil servants, and academics, ranging in age from 22 to 62 years. All participants had completed higher education in Afghanistan and possessed between two and 35 years of professional experience. At the time of the study, two participants were employed in Germany. Additionally, six women attended courses and language classes at Frankfurt UAS, aiming to qualify for employment in the social work sector. The participants had been living in Germany for between eight months to two years after their evacuation.

The FGDs adapted a set of questions originally developed for engaging with refugees, asylum seekers, and migrants in Europe (Kolendowicz 2017:52), and adhered to established guidelines for conducting such discussions (ETR 2013:4). Each session lasted between 1.5 and 2 hours and was structured into three main parts: background of the participants (including age, duration of stay, family, passage to Germany, socioeconomic status, education), experiences on arrival, and future plans. All discussions were conducted in Dari, recorded, transcribed, and partially translated for publication. A word count was adopted to determine the number of participants who engaged in a discussion and to identify key areas of concern.

8.4 Findings

8.4.1 Leaving Afghanistan

All participants reported that both their country and their families had experienced a period of significant progress prior to the Taliban's takeover in 2001. They highlighted their social status, contributions to society, and services to the people. Those who had worked in education emphasised their role in bringing modern education to the younger generation [G4-O].[1] Participants also spoke of their civic engagement in support of human rights and the common good. One woman, for example, shared that she had dedicated over 35 years to assisting victims of domestic violence [G1-Z].

Participants described the experience of leaving their country as profoundly disruptive and marked by hardship, yet also infused with hope. One activist and former politician expressed fear of retribution from the Taliban [G1-M], while others emphasised their determination to secure educational opportunities for their daughters [G3-W].

Fleeing to Germany was perilous. Kabul Airport was overcrowded, and transportation was delayed. One group of women witnessed a bomb explosion near the airport during their attempted evacuation [G1-Z]. Those who left later faced additional obstacles in obtaining the necessary documents, such as passports, visas, and the so-called No Objection Certificate from the Taliban [G3-T].

8.4.2 Arriving in Germany

Upon arrival in Germany, participants expressed appreciation for having immediate access to accommodation and financial support without the need to search for work [G2-F]. They also highlighted the positive aspect of being welcomed by social workers, local volunteers, and student 'buddies' from the STEPS-A programme, which helped ease their initial transition [G3-W, G4-J].

> "We live in an area where the shops are far away, but the social worker helped us by his car to get our groceries or go to our appointments" [G4-J].

Not all participants had a positive initial experience. One participant recounted that their family was forced to spend a night in the rain because the room provided was in a terrible condition, and the social worker refused to change it [G1-M].

[1] Participants-ID (G- Group, 4 is Number and Z First Letter of the Name)

Another family of five expressed that the refugee centre was unsuitable, as they were required to live together in a single room [G1–4].

> "When we came to Germany, we thought we were different from other refugees who came illegally, but later we realised we all are the same, and I felt zero with no identity" [G1-G].

The centres also posed other challenges:

> "I am facing challenges because of other Afghans who are in the centre, they want to control me and gossip about my lifestyle as in Afghanistan" [G2-F].

For younger women, going out alone was both a concern and a freedom. Feeling safe while shopping alone was a new experience for one woman at the age of 23 [G2-M]. Others experienced new freedoms as well:

> "Here I am living alone and going out anytime I want, but in Afghanistan, my father and brother always accompanied me in the evening when I went to university" [G2-N].

Another participant faced significant challenges, including experiences of social exclusion:

> "It was the 2nd day of our arrival, and we had a meeting at the Job centre; the moment I talked in English, the lady from the job centre told me that I was in Germany, and I had to talk in German. I felt extremely lost and did not know how one could expect us to speak a new language within two days" [G3-E].

All participants had to adapt to the complexity and rigidity of German bureaucracy:

> "The moment I open the post-box, I see a new letter and I feel stressed, then I have to find someone to read it for me and guide me: what does it mean?" [G1-M, similar G3-H].

Two groups highlighted that Afghans were covered by **health insurance** the moment they arrived and could see a doctor–if they managed to get an appointment [G1-G4]. One of the doctors compared the German healthcare system favourably to Afghanistan [G1-G].

8.4.3 Integrating in Germany

Learning a new language proved difficult, especially for the older generation. One woman even had to give up due to her health condition [G1-Z]. Some younger individuals claimed to speak German in less than a year [G3-S]. Others mentioned that it took them almost a year to find a place in the language and integration classes [G2-N, G4-J].

In the group of women with children and in the men's group, participants valued the fact that their children had the right to attend school without fees, particularly that girls could receive a good education in Germany, unlike in Afghanistan. One person noted that the children had enhanced freedoms in Germany and that Afghans had to become accustomed to different cultural norms [G3-W].

Those who were enrolled in the training programme at Frankfurt UAS appreciated this opportunity.

> "I am happy to continue my education based on my own choice here in Germany, while in Afghanistan my family forced me to study literature" [G2-M].

But the fear of not being able to use their existing knowledge and having to start afresh persisted [G1-M; G2]. The health professionals were also concerned that their training and experience would not qualify them to work in Germany [G1, G4]. In contrast, the male participants believed that individuals with expertise in information and technology (IT) and finance could begin working in Germany. Three of the six male participants had already found jobs in the country [G4].

Finding **accommodation** on the housing market proved to be challenging, with unknown rules and regulations applying due to the public funding of rent.

> "If you don't have a job, people don't rent their house to refugees" [G1-4].

Another challenge arose from the timing of the Afghan arrivals, which coincided with the influx of Ukrainian refugees. It appeared that the latter group may have been prioritised:

> "I was about to get an apartment, and on the day was to sign the contract, a Ukrainian family came to our centre, they gave the apartment to them and told me to wait" [G2-M].

8.4.4 Reuniting with Family

A major concern among participants is that their family members remain in Afghanistan, and they live in constant fear for their safety.

"We are a group of 35 young girls who came to Germany in evacuation, and our family is still in Afghanistan. I tried to bring my parents and siblings, but the German authorities told me that I am an adult, and I am not responsible for my family. We did a lot of advocacy, but got no results yet" [G2-1].

They had been instructed on the legal requirements for family reunification in Germany and found them nearly impossible to fulfil; no one was eligible to bring their family under the current admission programmes [G2-N, G1-Z]. Some of them had engaged in discussions with officials and advocated for their cases.

8.5 Discussion

This section summarises, interprets, and discusses the implications and limitations of the findings by contrasting applicable laws and policies with participants' narratives. A central point of reference is SDG Target 10.7, which aims to facilitate responsible migration through the implementation of well-managed migration policies—framed within the broader objective of **reducing inequality within and among countries** (Goal 10).

Germany's regulatory framework for migration is grounded in a combination of international, European, federal, and state laws, with implementation largely carried out at the local level. Important laws include the Residence Act (RA) and the Asylum Act (AA), supplemented by specific provisions concerning non-Germans in various other legal statutes.

The Migration Integration Policy Index for Germany ranks labour market integration and anti-discrimination efforts as 'favourable' (81 and 70 out of 100). Other factors are rated as only 'halfway favourable', including access to health (63/100), political participation (60/100), education (55/100), permanent residence (54/100), access to nationality (42/100), and family reunion (42/100).

Germany has opened its **labour market** to almost all migrants and refugees. However, professions such as medical and social work, along with many vocations, are highly regulated in terms of qualifications. To access the German labour market, holders of foreign certificates can apply for statements of 'Comparability' or 'Equivalence' through a Central Office. The recognition of foreign degrees has proven to be time-consuming. Furthermore, some professions—such as doctors and social workers—require a state licence, which necessitates advanced language skills, further training, and practical experience. Despite the demand for these professions ('Fachkräftelücke'), it remains difficult for trained and experienced professionals to access the respective jobs. One could observe ambition fading while waiting for documents and permissions.

Germany enacted a 'General Act on Equal Treatment' (AGG), which aims to **stop discrimination** on the grounds of race or ethnic origin, gender, religion or belief, disability, age, or sexual orientation (AGG, Sec. 1). This law has also led to the establishment of a network of local counselling centres. It highlights the need for action against discrimination rather than merely preventing it: many participants from various groups reported experiencing different forms of discrimination, but so far, no one has taken legal action or engaged in counselling.

While refugees and migrants are excluded from elections, the states have established 'Integration Commissions', which function as advisory boards to the local government, allowing 'foreigners' to have a voice. The German government has consulted with representatives regarding admission programmes. Consequently, female Afghan evacuees were involved early on in discussions about the definitions of 'household' and 'family member' in the draft admission programme. Members of training courses also began discussing the establishment of their own association.

Humanitarian visas ensure unemployment benefits, health insurance, and access to education and work from day one. New arrivals are accommodated in special reception centres where residency requirements can be imposed. All groups found the educational and health systems 'favourable'. In theory, families were entitled to childcare; in practice, however, places were lacking. This negatively affects children with migration backgrounds, as institutionalised childcare helps to reduce educational inequalities. Moreover, some female participants with small children were forced to revert to a 'traditional' division of labour within the family (which they did not experience in Afghanistan). Compulsory schooling—in some states like Hesse—up to age 21 and free education were considered significant advantages for families with girls who could not have attended school if they had remained in Afghanistan. The Afghan evacuees appear to be very ambitious regarding education, with many starting early and working hard to achieve the necessary language proficiency to enter university (which is open to humanitarian refugees).

To enter and stay in Germany, migrants require a residence title (RA, Sec. 4). Afghans benefit from a special provision in the German Residence Law (RA) that allows the Federal Ministry of the Interior to grant residence 'for reasons of international law or on urgent humanitarian grounds' (RA, Sec. 22). The provision that subsistence must be secured without recourse to public funds (RA, Sec. 2) does not apply to humanitarian visas. Permanent residency and citizenship depend on this provision (RA, Sec. 9). This means Afghan refugees are safe as long as their country remains unsafe. Meanwhile, they can progress and apply for citizenship if they wish. Still, Afghan participants expressed fear due to their temporary status, especially when discussions arose about the deportation of—usually—young single Afghan men (in one instance following a lethal knife attack on a police officer). In one FGD [G2], a participant referenced an alleged regulation in France that grants a seven-year residence permit, in contrast to the two-year permission in Germany. This remark offers insight into how information—whether accurate or not—circulates within communities in exile.

Holders of a residency permit in Germany are permitted to bring immediate family members for the purpose of 'protecting marriage and the family'. However, this right is restricted for individuals who rely on social benefits (RA, Sec. 27). The legal definition of family in this context is limited to spouses and underage children; other relatives may be admitted only in cases of exceptional hardship (RA, Sec. 36). In 2023, the state of Hesse introduced an admission programme for Afghan family members, which initially raised hopes among the Afghan community. However, the programme presented significant and often insurmountable barriers—for example, while a brother might be included, his wife could be excluded, and sponsors were required to provide financial guarantees amounting to €50,000 per person. These conditions left many in the community feeling disappointed and discouraged.

8.6 Recommendations for Social Interventions

For the group of Afghan evacuees, support from student buddies and social workers at the reception centres proved valuable during their initial resettlement. However, the integration process could have been further strengthened by the presence of trained counsellors—both from within the Afghan community and from the local population—who could provide culturally sensitive guidance to the 'newcomers'.

Additionally, greater care should be taken to provide 'newcomers' with easily understandable information that takes into account their cultural and linguistic backgrounds.

Access to employment is crucial for fostering self-esteem and a sense of agency among refugees. While language acquisition is important, insisting on full proficiency before allowing individuals to work can be demotivating and counterproductive. The professional and cultural expertise of those who were forced to flee—many of whom had already established meaningful careers in their home country—should be recognised and utilised, rather than allowed to go to waste.

Legal and political adjustments are necessary to reflect more inclusive understandings of family and kinship. Host societies should recognise that the concept of 'family' often extends beyond the nuclear model envisioned by the German Constitution. A sense of responsibility toward extended family members does not cease at national borders. Creating realistic opportunities for individuals to care for their relatives can serve as a powerful incentive for integration.

Policies and strategies should be developed *with* Afghan refugees, not *for* them Inclusive, participatory approaches foster a sense of belonging and agency—both of which, as Hannah Arendt reminds us, are often lost by those who have been forced to flee their country.

References

Arendt H (1943) We Refugees Menorah Journal 31:69–77

Berlin DIW (2023) Employment opportunities of refugee women in Germany are improving despite starting at a disadvantage. DIW Wkly Rep 19+20):130–136

Böhm L (2019) Geflüchtete Frauen und Familien: Literaturüberblick zu gender- und familienbezogenen Aspekten der Fluchtmigration und Integration. Institut für Arbeitsmarkt- und Berufsforschung, Nuremberg

Brücker H, Kunert A, Mangold U et al (2016) Geflüchtete Menschen in Deutschland - eine qualitative Befragung. Institut für Arbeitsmarkt- und Berufsforschung, Nuremberg

Bundesamt für Migration und Flüchtlinge (BAMF) (2021a) Female refugees in Germany—ecreational habits and social contacts. Bundesamt für Migration und Flüchtlinge, Nuremberg

Bundesamt für Migration und Flüchtlinge (BAMF) (2021b) Kurzanalyse: Spracherwerb und soziale Kontakte schreiten bei Geflüchteten voran. Bundesamt für Migration und Flüchtlinge, Nuremberg

Currle E (2011) Focus group discussions in migration research: New directions in migration. https://www.researchgate.net/publication/283279052. Accessed 18 Mar 2025

ETR (2013) Best practices in research & evaluation: focus groups. ETR Best Practice Guides, Scotts Valle

Kolendowicz E (2017) Focus groups report with refugees, asylum seekers and migrants. https://www.workexperienceinlondon.com/images/refuges/FGR2.pdf. Accessed 18 Mar 2025

Matsangos M, Ziaka L, Exadaktylos AK et al (2022) Health status of afghan refugees in Europe: policy and practice implications for an optimised healthcare. Int J Environ Res Public Health 19(15):9157. https://doi.org/10.3390/ijerph19159157

Nylen AJ, Omar B, Bott J et al (2023) 'Then, we lost everything': afghan evacuee experiences of operation allies refugee and operation allies welcome. Brown University, Providence

Safi AQ, Agustin H, Rizal E (2023) Afghan migrants in Germany: cross-cultural communication and impact of immigration on afghan culture. Stud Media Commun 11(7):135

Safi AA, Czaika M (2024) The transnational engagement of afghan diaspora organizations: drivers of diaspora specialization. Glob Netw 25(1):e12484. https://doi.org/10.1111/glob.12484

Sharifian F, Sadeghpour M, Barton S-M et al (2020) English language learning barriers of afghan refugee women in Australia. Int J Appl Linguist 31:65–78

Stempel C, Alemi Q (2021) Challenges to the economic integration of afghan refugees in the U.S. J Ethn Migr Stud 47(21):4872–4892

Swedberg R (2020) Exploratory research. In: Elman C, Gerring J, Mahoney J (eds) The production of knowledge: enhancing progress in social science. Cambridge University Press, Cambridge, pp 17–41

9 Addressing the Legal Void: Protecting Rohingya Refugee Children in Malaysia Through Social Work Intervention

Pavithra Sukumaran, Mohd Suhaimi Mohamad, and Noremy Md Akhir

Abstract

Malaysia hosts a considerable influx of Rohingya refugees from Myanmar each year. Despite the country's international commitments, the lack of specific legal measures within domestic laws exposes Rohingya refugees to exploitation and denies them fundamental protection mechanisms. This includes Rohingya children, who are vulnerable to various forms of exploitation such as trafficking, child labour, child marriage, and an increased risk of arrest or detention. Utilising qualitative research with a descriptive phenomenological approach, this chapter explores the vulnerabilities faced by Rohingya refugee children in Malaysia due to the absence of a legal framework, in line with the Sustainable Development Goals (SDGs) 10.3 and 10.7. Through in-depth interviews with eight informants aged 12 to 17 from Perak, Malaysia, this chapter underscores the urgent need for social work intervention. Social workers can provide individualised support, advocacy, and collaboration with stakeholders to safeguard the rights and well-being of these vulnerable children. By addressing systemic gaps and promoting community empowerment, social workers can create a safer and more inclusive environment for refugee children to thrive and realise their full potential. This research strongly urges the prioritisation of the welfare and respect of Rohingya refugee children, stressing the critical need to address their protection requirements within Malaysia's policy framework.

Keywords

Rohingya refugee children · Descriptive phenomenological approach · Exploitation and vulnerabilities · Legal framework · Social work intervention · Sustainable Development Goals (SDG 10.3 and 10.7)

P. Sukumaran (✉) · N. M. Akhir
Faculty of Social Sciences and Humanities, The National University of Malaysia (UKM), Bangi, Malaysia
e-mail: pavithrapavitu@gmail.com

N. M. Akhir
e-mail: noremy@ukm.edu.my

M. S. Mohamad
Industry and Community Partnership, Faculty of Social Sciences and Humanities, The National University of Malaysia (UKM), Bangi, Malaysia
e-mail: msuhaimi@ukm.edu.my

9.1 Introduction

Global migration is a complex phenomenon involving the movement of people across national borders, either voluntarily or involuntarily, often resulting in temporary or long-term residence in a host country. Involuntary migration, which frequently occurs due to conflict or natural disas-

N. V. Roman et al. (eds.), *Social Inequalities Within and Among Countries*, Sustainable Development Goals Series, https://doi.org/10.1007/978-3-032-08741-6_9

ters, happens when individuals are compelled to move against their will (European Parliament 2024). Southeast Asia functions as a major migration hub, comprising countries of origin, destination, and transit (Migration Data Portal 2020). In the context of Malaysia, major migration often involves the influx of refugees.

According to the 1951 Convention and its 1967 Protocol, a refugee is defined as a person who has fled their country due to a well-founded fear of persecution based on race, religion, nationality, membership in a particular social group, or political opinion, and who is unable or, because of such fear, unwilling to seek protection from their home country (European Union and United Nations, 2018). Although Malaysia is not a signatory to the 1951 Refugee Convention (Supaat 2025), it hosts a substantial refugee and asylum-seeking population. As of end January 2025, 192,420 individuals were registered with the UNHCR in Malaysia, including a significant number of Rohingya refugees, who alone accounted for 111,980 (UNHCR 2025).

Rohingya refugees are a predominantly Muslim minority group residing in Myanmar's Rakhine State. Owing to decades of systemic persecution in Myanmar, Rohingya individuals have been seeking refuge in Malaysia since the 1990s (Letchamanan 2013; Yesmin 2016). As a key host country for displaced Rohingya populations, Malaysia is home to a substantial and growing Rohingya refugee community (Shaw et al. 2019).

Children are disproportionately represented among refugee populations. In 2020, the number of international migrant children worldwide reached an estimated 35.5 million—the highest figure ever recorded (United Nations Children's Fund [UNICEF] 2024). Among the Association of Southeast Asian Nations (ASEAN), Malaysia hosts the second-largest population of child refugees and asylum seekers (Migration Data Portal 2020). As of end January 2025, 54,340 Rohingya refugee children were registered with the UNHCR in Malaysia (UNHCR 2025), a figure that excludes undocumented children.

In Malaysia, refugee children face significant challenges due to their lack of legal status and the government's perception of them as a transient community with limited rights for long-term stay (Asia Pacific Refugee Rights Network 2019). The absence of a legal framework for protection results in refugee children being denied their fundamental rights, such as (UNHCR 2019). This situation is further exacerbated for undocumented refugees who do not fall under the UNHCR's protection (Farzana et al. 2020). The challenges faced by Rohingya refugee children in Malaysia are severe and complex, highlighting the urgent need for a domestic legislative and administrative framework to safeguard their rights (UNICEF 2022). However, the current legal structure does not extend protection to Rohingya refugee children or adequately address the vulnerabilities they face. In the absence of specific legal provisions for safeguarding children, those residing in Perak—and across Malaysia—are at heightened risk.

9.2 Literature Review

9.2.1 Lack of a Legal Framework for Refugees in Malaysia

Since Malaysia is not a signatory to the 1951 Refugee Convention (United Nations 1951), its domestic legal framework neither formally recognises nor differentiates undocumented refugees. Instead, national legislation criminalises their presence, leading to the absence of a clear legal definition of 'refugee' within the Malaysian context. Although Malaysia has ratified the United Nations Convention on the Rights of the Child (CRC), which obligates states to ensure the protection and humanitarian assistance of refugee children, it maintains reservations on several key provisions that are particularly important for safeguarding the rights of refugee children.[1]

[1] An important Article is Article 22. "Malaysia is a party to the UN Convention on the Rights of the Child (CRC) which obligates States to ensure refugee children have the right to receive protection and humanitarian assistance". However, Malaysia still holds reservations towards Article 2 (regarding non-discrimination); Article 7 (regarding birth registration, the right to a name and nationality); Article 14 (regarding freedom of thought, conscience and religion); Article 28(1) (a) (regarding compulsory and free primary

In addition, Section 6(1)[2] of the Immigration Act of 1959/63 prohibits the entry of non-citizens into Malaysia, while Section 6(3)[3] criminalises and penalises undocumented individuals, including refugees who are awaiting documentation from the UNHCR (Jabatan Imigresen Malaysia 2022). This legal stance stands in contradiction to Article 8 of the Federal Constitution—specifically Article 8(1)—which affirms that all persons are equal before the law and entitled to equal protection of the law), a provision that extends to refugees as well. Moreover, Malaysia's Child Act 2001 affirms that every child is entitled to protection and assistance in all circumstances, without discrimination based on race, colour, sex, language, religion, social origin, physical, mental or emotional disability, or any other status.[4] Therefore, this significant deficiency in the legal framework contributes to the vulnerable situation faced by Rohingya refugee children in Malaysia.

9.2.2 Vulnerabilities due to the Lack of a Legal Framework

The absence of a legal framework (UNICEF 2022) and official refugee status in Malaysia exposes refugees to various forms of abuse and exploitation in the workplace, including non-payment or partial payment of wages, verbal abuse, arbitrary dismissal, physical violence, sexual harassment, and workplace raids (Smith 2012).

Refugee children, asylum seekers, and those in irregular migration situations face the constant threat of detention, which places them at heightened risk of exploitation and abuse within immigration detention facilities. In 2016, immigration detention centres (IDCs) held 2290 children who were either refugees or asylum seekers (Parthiban and Khoo 2019). In October 2020, the Home Ministry reported that 756 children were being held in IDCs (US Department of State 2020). Currently, there are 14 operational IDCs in the country (International Detention Coalition 2020), none of which provide separate accommodations for children. As a result, children are detained alongside adults, exposing them to reported instances of 'appalling' conditions, physical violence by officers (Save the Children 2017), potential abuse from adult detainees, and significant deterioration in both physical and mental health (Child Rights Coalition Malaysia 2020). In 2022 alone, 150 deaths were recorded in IDCs, including seven child fatalities.[5]

On the other hand, Rohingya refugee children face trafficking, child marriage, exploitation, and abuse (Hutchinson 2018). In Malaysia and around the world, efforts to eradicate child marriage is gaining momentum. According to UNICEF (2022), approximately 15,000 cases of child marriage were recorded in Malaysia between 2007 and 2017, spanning all communities, including the Rohingya community. Child marriage undeniably deprives children of the opportunity to grow, learn, and reach their full potential, while exposing them to risks such as abuse, early pregnancy, and long-term harm to their well-being. Fortify Rights (2019) documented a case of a 16-year-old Rohingya refugee girl in Kuala Lumpur who expressed deep dissatisfaction with her early marriage, describing experiences of both physical

education for all); and Article 37 (regarding torture or other cruel, inhuman or degrading treatment or punishment and unlawful or arbitrary deprivation of liberty).

[2]Malaysia. (1 January 2006). Immigration Act 1959/63, Section 6 (1). "No person other than a citizen shall enter Malaysia unless –(a) he is in possession of a valid Entry Permit lawfully issued to him under Section 10; (b) his name is endorsed upon a valid Entry Permit in accordance with Section 12, and he is in the company of the holder of the Permit; (c) he is in possession of a valid Pass lawfully issued to him to enter Malaysia; or (d) he is exempted from this section by an order made under Section 55". Available: https://www.imi.gov.my/index.php/pekeliling/immigration-act-1959-63-act-155/

[3]Ibid. Section 6 (3). "Any person who contravenes subsection (1) shall be guilty of an offence and shall, on conviction, be liable to a fine not exceeding ten thousand ringgit or to imprisonment for a term not exceeding five years or to both and shall also be liable to whipping of not more than six strokes".

[4]Child Act 2001. Act 611. [1 August 2002, P.U. (B) 229/2002], p.13. Retrieved from https://antislaverylaw.ac.uk/wp-content/uploads/2019/08/Malaysia-Child-Act-1.pdf

[5]Malaysia Now. 23 February 2023. 'Government Pressed to Probe Deaths of 150 Foreigners in Detention Last Year'. www.malaysianow.com/news/2023/02/23/govt-pressed-to-probe-deaths-of-150-foreigners-in-detention-last-year

and verbal abuse from her spouse. Currently, there are no proper protection mechanisms in place to identify and support victims of domestic violence linked to child marriage, and such cases are rarely reported.

Despite seeking safety and refuge in Malaysia, Rohingya refugee children often become entangled in complex human trafficking networks. The Rohingya population from Myanmar has been particularly vulnerable to trafficking in Malaysia—a concern highlighted by Maria Grazia Giammarinaro, the United Nations (UN) Special Rapporteur on trafficking in persons, especially women and children (United Nations Human Rights Council [UNHRC] 2015). The report also highlighted the prevalence of child trafficking for the purpose of forced begging[6] within the refugee population, a practice that predominantly affects Rohingya refugee children.

Child labour directly contravenes the principles outlined in the CRC (Dandavati and Tan 2019). A 2019 report by the Special Rapporteur on the sale and sexual exploitation of children highlights that Malaysia continues to be a destination for trafficking, forced marriage, and sexual exploitation, driven in part by high demand for such exploitation and the forced begging of refugee women and children (United Nation 2019). Additionally, the report reveals that 101 refugee children were found by UNHCR to be involved in begging and trapped in debt bondage. Further research conducted by Abdullah et al. (2018) documented the case of a 12-year-old Rohingya refugee girl who resorted to begging and selling religious books in order to survive. Given the precarious circumstances and limited rights of these refugee children, it is evident that their lives in Malaysia are challenging. This underscores the urgent need for child protection interventions to address the vulnerabilities of refugee children living in Malaysia.

The intersectionalities of Rohingya refugee children in Malaysia—including their refugee and migration status, ethnicity, nationality, economic class, gender, and age—create a complex web of overlapping vulnerabilities. These children often lack legal status and documentation, face the constant threat of detention, and are exposed to heightened risks of trafficking and child labour. In addition, they are affected by the prevalence of child marriage, poor health and well-being, economic hardships, limited access to education, social marginalisation, and gender-based vulnerabilities. Collectively, these factors exacerbate their precarious living conditions and highlight the urgent need for comprehensive child protection interventions.

9.3 Methodology

This study employed a qualitative research design to investigate real-world problems through semi-structured interviews (Moser and Korstjens 2017). A phenomenological approach was adopted to explore the participants' 'lived experiences', aiming to uncover the underlying reasons for their behaviours and to gain insight into their perspectives (Alhazmi and Kaufmann 2022). Purposive sampling was used to identify and select informants based on clearly defined criteria relevant to the study's objectives. The study involved eight participants aged 12 to 17 years who are Rohingya refugee children. A strong rapport was established with the selected participants through close collaboration. Each interview session lasted an average of 45 to 60 minutes over ten sessions. The collected data were analysed using a descriptive phenomenological approach, which involved a systematic process of identifying and describing the lived experiences of participants (Giorgi et al. 2017). This included immersing in the data to gain a deep understanding, extracting significant statements, formulating meanings, clustering these meanings into themes, and synthesising the essence of the experiences. NVivo was used as a

[6]Malaysia. Child Act 2001. Act 611, Section 32 (a). "Any person who causes or procures any child or, being a person having the care of a child, allows that child to be on any street, premises or place for the purposes of begging, receiving alms, whether or not there is any pretense of singing, playing, performing or offering anything for sale, commits an offense and shall on conviction be liable to a fine not exceeding five thousand ringgit or to imprisonment for a term not exceeding two years or to both" (p. 43). https://antislaverylaw.ac.uk/wp-content/uploads/2019/08/Malaysia-Child-Act-1.pdf

tool to organise and manage the data during the analysis process, supporting the identification of emerging meanings while ensuring the integrity of participants' narratives was maintained.

9.4 Ethical Considerations

Ethical approval for this research was obtained from the National University of Malaysia (UKM). Participants were provided with detailed information sheets outlining the purpose and scope of the study and gave signed consent for both participation and audio recording. To ensure confidentiality and protect the identities of Rohingya refugee children in Perak, Malaysia, all audio recordings were anonymised and identifying information was removed from research documents.

9.5 Research Site

The study was conducted in Menglembu, in the Kinta District of Perak, Malaysia. The researcher chose Menglembu due to prior involvement in social work initiatives targeting this refugee community, as well as collaborative efforts with local refugee organisations.

9.6 Results and Discussion

This section presents a thematic analysis of the multidimensional protection challenges experienced by Rohingya refugee children in Malaysia, as revealed through their lived narratives. It critically examines four key areas: vulnerability to law enforcement through arrests and detention, exposure to child trafficking, the prevalence of child labour, and the persistence of child marriage. These sub-themes expose how the lack of legal recognition, weak enforcement of international and domestic child protection standards, and deep-rooted socio-economic inequalities systematically marginalise Rohingya refugee children. This analysis is directly aligned with SDG 10.3, which calls for the elimination of discriminatory laws and practices and the promotion of equal opportunity, as well as SDG 10.7, which advocates for safe and responsible migration through well-managed policies.

9.6.1 Vulnerability to Law Enforcement: Arrests and Detention

The findings underscore a prevalent concern about the increased vulnerability of Rohingya refugee children to arrests and subsequent detention by Malaysian authorities. Due to the lack of legal documentation and UNHCR cards, these children and their families are at constant risk of detention by the police. One informant provided evidence detailing their experiences of arrest and detention due to the lack of legal documentation, as follows:

> "In 2018, I was detained and locked up in a police station for 5 days together with my parents. Yes, I have no UNCHR card, but I have already applied. My parents informed the police officer about my application and even showed my application paper. But they refused to listen to them. We don't know to whom we should ask for help. We are helpless because this is not our country. We still face fear, wherever we go, there is this catch"

(Informant 1, 16 years old).

The findings clearly demonstrate that undocumented Rohingya refugee children in Malaysia face acute vulnerability to law enforcement, living in constant fear of arrest and detention. This situation is particularly concerning given that Malaysia has ratified the CRC, which mandates protection and humanitarian assistance to refugee children. However, Malaysia's non-signatory status to the 1951 Refugee Convention results in these children being denied the rights afforded to Malaysian citizen children. This, combined with discriminatory legal frameworks and the non-recognition of Rohingya ethnicity, places these children in a state of extreme precarity. Furthermore, the findings highlight a glaring lack of institutional support for this already marginalised group, underscoring the urgent need for targeted child protection interventions.

9.6.2 Child Trafficking

While Malaysia has ratified the Convention on the Elimination of All Forms of Discrimination Against Women (CEDAW), a fundamental international accord safeguarding women and girls, Rohingya refugee children still face an increased risk of trafficking due to their lack of legal recognition in Malaysia. These findings underscore the vulnerability of these children to trafficking, as their absence of legal status and protection, coupled with weak legal frameworks and enforcement mechanisms, leaves them exposed to exploitation. This reality is supported by the testimony of one informant:

> "Sadly speaking, our situation is beyond the worst. Women and girls, especially children like us, often fall victim to human trafficking, particularly in Thailand. My sister aged 18, become one of the victims of this. It's been a year since I didn't meet her. My parents couldn't even lodge a police report. Because we're refugees here. We don't have a complaint mechanism to report our abuse. We face this in Myanmar, yet in Malaysia, also same situation"

(Informant 4, 16 years old).

This testimony from a 16-year-old Rohingya refugee powerfully illustrates the dire circumstances faced by Rohingya children in Malaysia. Despite the country's endorsement of key legal instruments such as the CRC and the Child Act 2001, a significant gap remains between legal commitments and actual implementation. This gap leaves Rohingya refugee children—already deprived of legal status—at heightened risk of trafficking and exploitation. Without formal recognition or legal protection, they are effectively powerless to seek help or report abuse, rendering them vulnerable not only in their host country but also to the same threats they fled in Myanmar.

9.6.3 Child Labour

There is a growing concern among the Rohingya community in Malaysia regarding the widespread involvement of their children in child labour. The lack of access to formal education significantly exacerbates the problem, as Rohingya refugee children are barred from attending public schools and are often compelled to work to support their families. Testimonies from two informants further substantiate this issue, confirming the harsh realities these children face.

> "I'm just 14 years old, but I ended up working in the market, which is pasar borong and pasar malam with my parents. Each time, I see other children going to school … I feel like want to wear school bag and join them and study in school. But what can I do, I'm just a refugee in this country, so I cannot go to public school … my ambition gone"

(Informant 2).

> "I'm working as a mechanic boy in a car workshop near my house. My owner paid only RM 30.00 daily … he will give me many work. Sometimes, I don't have time to rest. If I work a bit slowly, my owner will scold me in front of others. I must work 6 days a week, feel so tired. If I'm sick, he will ask me to come to work"

(Informant 5, 16 years old).

The testimonies of Rohingya refugee children in Malaysia highlight the urgent issue of child labour. Despite international conventions such as the CRC and domestic legislation like the Child Act 2001, which state that children under a certain age should not be engaged in labour, these children are forced into exploitative work environments (Mixed Migration Centre 2023).

The poor implementation of the CRC and the Child Act 2001 increases the vulnerability of refugee children to child labour. Discriminatory laws and policies, combined with their lack of legal status, make them particularly susceptible to exploitation. While laws prohibiting child labour exist, enforcement mechanisms are often inadequate, allowing unscrupulous employers to exploit vulnerable children.

9.6.4 Child Marriage

Child marriages frequently take place without the consent of the children involved. Among Rohingya refugee children in Malaysia, this practice remains prevalent, driven by economic hardships and the lack of legal protections. Families with little or no income resort to arranging marriages for their young daughters as a survival strategy. Such marriages deprive children of their rights to education, health, and personal development, and are frequently associated with early pregnancies, domestic violence, and long-term psychological trauma. One informant further affirmed the severity of this issue, stating:

> "I'm just 16 years old. My parents forced me to marry a Rohingya man. They said if I marry him, I will have a better life. But I want to be a scientist, that is my ambition. I want to study. But my parents didn't value my ambition and married me to him. The sad part is my husband never let me go out. I stayed in the room alone all day. I feel like lifeless being like this"

(Informant 4[7]).

The testimony of this 16-year-old Rohingya refugee in Malaysia starkly illustrates the challenges faced by many in similar circumstances. Although Malaysia has withdrawn its reservation to Article 16(2) of the CEDAW—which prohibits child betrothal and marriage—the country lacks specific legal measures to prevent child marriage. The continued prevalence of this practice among Rohingya refugee children in Malaysia is particularly concerning. It underscores the reality that those subjected to child marriage are often denied the opportunity to pursue their aspirations and experience the deep distress of social isolation (Guglielmi et al. 2021).

9.7 Proposed Social Work Intervention

The lack of recognition and protection for Rohingya refugee children in Malaysia significantly impacts SDG 10, particularly Targets 10.3 and 10.7, which focus on reducing inequalities and facilitating safe, orderly, and responsible migration and mobility. SDG 10.3 aims to ensure equal opportunity and reduce inequalities by eliminating discriminatory laws, policies, and practices. The marginalisation and vulnerability of Rohingya refugee children due to the lack of legal recognition highlight the need to address systemic inequalities. Social work interventions can play a crucial role by advocating for policy changes, providing direct support, and working to eliminate discriminatory practices. Meanwhile, SDG 10.7 focuses on facilitating safe and responsible migration through well-managed policies. The precarious status of Rohingya refugees underscores the urgent need for comprehensive migration policies that protect their rights. Social workers can assist in navigating asylum processes, providing legal aid, and supporting resettlement efforts, thereby contributing to safer migration practices.

Concerning the above findings, social work interventions are essential. At the micro level, social workers can provide individualised support and advocacy for Rohingya refugee children by offering legal assistance to navigate immigration and asylum processes. Additionally, they can facilitate access to trauma-informed counselling and support services to address the psychological impact of trafficking experiences. Social workers can collaborate with educational and vocational training centres to design programmes that equip refugee children with skills for sustainable livelihoods and assist in resettlement programmes. Through community engagement, social workers can challenge harmful practices like child marriage and promote gender equality, educating parents about the negative impacts of child marriage and the importance of child welfare. Moreover, social workers can advocate for inclusive policies that protect the rights of refugee children and eliminate discriminatory practices.

By addressing these issues, social workers can mitigate the vulnerabilities faced by Rohingya refugee children, thereby contributing to the reduction of social inequalities (SDG 10.3) and promoting safe and responsible migration (SDG 10.7). Through targeted interventions, social workers can empower these children to build brighter futures and integrate more fully into society, ultimately narrowing inequality gaps and providing better opportunities for a stable and secure life.

[7]The researcher interviewed the respondent in her room, with the consent of her husband, after explaining the purpose of the research. The collaboration with refugee organisations in Menglembu facilitated this process.

9.8 Conclusion

In conclusion, the chapter underscores the vulnerabilities confronting Rohingya refugee children in Malaysia due to the absence of a legal framework, in alignment with SDG 10.3 and 10.7. Despite Malaysia's significant refugee population, the lack of specific legislation exposes refugee children to various forms of exploitation, including trafficking, child labour, and child marriage, and makes them more prone to arrest or detention. The chapter highlights the urgent need for social work intervention to address these challenges. Through individualised support, advocacy, and collaboration with relevant stakeholders, social workers can play a crucial role in safeguarding the rights and well-being of Rohingya refugee children. By addressing systemic gaps and promoting community empowerment, social workers can help create a safer and more inclusive environment for refugee children to thrive and realise their full potential.

References

Abdullah A, Mohamed Dali A, Abd Razak MR (2018) Surviving stateless refugees: the uncertain future of Rohingya's children in Malaysia. Polit & Secur Res Cent, UKM 45(1):177

Alhazmi AA, Kaufmann A (2022) Phenomenological qualitative methods applied to the analysis of cross-cultural experience in novel educational social contexts. Front Psychol 13:785134. https://doi.org/10.3389/fpsyg.2022.785134

Child Rights Coalition Malaysia (2020) Status report on child rights in Malaysia 2019. Petaling Jaya, Child Rights Coalition Malaysia, p 81

Dandavati N & Tan S (2019) Where's the child? The rights of child domestic violence survivors. Women's Aid Organisation. https://wao.org.my/wp-content/uploads/2020/12/WAO-Strengthening-the-Primary-Healthcare-Response-toDomestic-Violence.pdf

European Union and United Nations (2018) International recommendations on refugee statistics (Expert Group on Refugee and Internally Displaced Persons Statistics). https://unstats.un.org/unsd/demographic-social/Standards-and-Methods/files/Principles_and_Recommendations/International-Migration/2018_1746_EN_08-E.pdf

European Parliament (2024) Exploring migration causes: why people migrate. https://wwweuroparleuropaeu/topics/en/article/20200624STO81906/exploring-migration-causes-why-people-migrate. Accessed 18 Mar 2025

Farzana A, Pero SD, Othman MF (2020) The dream's door: educational marginalization of Rohingya children in Malaysia. South Asian J Bus Manag Cases 9(2):237–246

Fortify Rights (2019) Malaysia: end child marriage, protect Rohingya refugee girls https://wwwfortifyrightsorg/mly-inv-jnr-2019-02-21/ Accessed 18 Mar 2025

Giorgi A, Giorgi B, Morley J (2017) The descriptive phenomenological psychological method. In: Willig C, Stainton-Rogers W (eds) The SAGE handbook of qualitative research in psychology. SAGE Publications, pp 176–192. https://www.researchgate.net/publication/318451180_The_Descriptive_Phenomenological_Psychological_Method

Guglielmi S, Mitu K, Seager J (2021) "I just keep quiet": addressing the challenges of married Rohingya girls and creating opportunities for change. Eur J Dev Res 33(6):1232–1251. https://doi.org/10.1057/s41287-021-00437-6

Hutchinson S (2018) Gendered insecurity in the Rohingya crisis. Aust J Int Aff 72(1):1–9. https://doi.org/10.1080/10357718.2017.1402291

International Detention Coalition (2020) Strengthening alternatives to immigration detention for children: mapping and assessment of residential care centres in peninsular Malaysia. Melbourne, International Detention Coalition, p 6

Jabatan Imigresen Malaysia (2022) Immigration Act 1959/63 [Act 155]. www.imi.gov.my/index.php/pekeliling/immigration-act-1959-63-act-155/

Letchamanan 2013 H (2013) Myanmar's Rohingya refugees in Malaysia: education and the way forward. J Int Comp Educ 2(2):86–97https://doi.org/10.14425/00.50.24

Malaysia Now (2023) Government pressed to probe deaths of 150 foreigners in detention last year. https://www.malaysianow.com/news/2023/02/23/govt-pressed-to-probe-deaths-of-150-foreigners-in-detention-last-year

Migration Data Portal (2020) Migration data in South-eastern Asia. https://wwwmigrationdataportalorg/regional-data-overview/south-easternasia Accessed 12 Dec 2022

Mixed Migration Centre (2023) Children and youth in mixed migration. http://www.mixedmigration.org/

Moser A, Korstjens I (2017) Series: practical guidance to qualitative research. Part 1: introduction. Eur J Gen Pract 23(1):271–273

Parthiban S, Khoo YH (2019) Detention of refugee children in Malaysia and Thailand. J Southeast Asian Hum Rights 3(1):59. Retrieved from https://doi.org/10.19184/jseahr.v3i1.10142

Save the Children (2017) Unlocking childhood: current immigration detention practices and alternatives for child asylum seekers and refugees in Asia and the Pacific. Save the Children Australia, Melbourne, p 37

Shaw SA, Karim H, Bellows N et al (2019) Emotional distress among Rohingya refugees in Malaysia. Intervention 17(2):174–180

Smith A (2012) In search of survival and sanctuary in the city: refugees from Myanmar/Burma in Kuala Lumpur. International Rescue Committee, New York, Malaysia

Supaat D (2025) Building bridges, not walls: the case for refugee family integration in Malaysia. Int J Stud Child Women Elder Disabl 23. https://ijcwed.com/wp-content/uploads/2025/02/IJC23-044.pdf

UNHCR (2025) Figures at a glance in Malaysia. https://www.unhcr.org/my/what-we-do/figures-glance-malaysia

UNICEF (2024) International Data Alliance for Children on the Move. https://data.unicef.org/resources/international-data-alliance-for-children-on-the-move/#

United Nations Children's Fund (UNICEF) (2022) The impact of COVID-19 on access to education and healthcare for refugees and asylum-seeking children in Peninsular Malaysia. https://www.unicef.org/malaysia/media/3006/file/Left%20Far%20Behind%20%20Full%20Report.pdf. Accessed 18 Mar 2025

United Nations (2019) Report of the Special Rapporteur on the sale and sexual exploitation of children, including child prostitution, child pornography and other child sexual abuse material (A/HRC/40/51/Add.3). https://docs.un.org/en/A/HRC/40/51/Add.3

United Nations High Commissioner for Refugees (UNHCR) (2018) Bangladesh, Operational update, 19–31 July https://data2unhcrorg/en/documents/download/65038 Accessed 18 Mar 2025

United Nations Human Rights Council (UNHRC) (2015) Report of the special rapporteur on trafficking in persons, especially women and children, Maria Grazia Giammarinaro https://www.refworld.org/docid/5577e9f34.html. Accessed 18 Mar 2025

United Nations (1951) Convention relating to the status of refugees. United Nations Treaty Series, 189, 137. https://treaties.un.org/doc/Treaties/1954/04/19540422%2000-23%20AM/Ch_V_2p.pdf

U.S. Department of State (2020) 2020 country reports on human rights practices: Malaysia. https://www.state.gov/reports/2020-country-reports-on-human-rights-practices/malaysia/

Yesmin S (2016) Policy towards Rohingya refugees: a comparative analysis of Bangladesh, Malaysia and Thailand. J Asiat Soc Bangladesh 61:71–100

10 'Difficult Life' and 'Pushing Life': Poverty and Vulnerability in Female-Headed Households in a Peri-Urban Township of Evaton, South Africa

Motlalepule Nathane

Abstract

Globally, historical inequalities have been identified as major obstacles to sustainable development, frustrating countries' efforts to eradicate poverty. These inequalities have also constrained opportunities for populations, particularly women, to participate meaningfully in social, cultural, political, and economic life. South Africa has grappled with the triple challenge of poverty, inequality, and unemployment both before and after the democratic transition of 1994. These social issues have been recognised as threats to democracy and as underlying causes of community protest and political unrest. In South Africa, inequalities are experienced both at the household level and along racial and gender lines. This chapter is based on a qualitative study that utilised a narrative design with 15 female-headed households in Evaton, a peri-urban township. The findings indicate that these households face severely limited livelihood opportunities, condemning them to a persistent state of poverty, which participants described as living a 'difficult life'. The study concludes that the combination of economically disadvantaged backgrounds and limited income-generation opportunities subjects female-headed households to ongoing poverty and vulnerability. The study recommends that the state implement meaningful interventions to sustain livelihoods and expand employment and income opportunities for a broader spectrum of society, particularly through the creation and support of local industries.

Keywords

Evaton township · Female headed households · Feminization of poverty · Peri-urban · Sustainable development · Unemployment and vulnerability

10.1 Introduction

South Africa has been identified as a country with the highest levels of inequality in the world, ranking first among 164 countries in the global poverty database (Sulla et al. 2022). To understand these inequalities, it is necessary to consider the contextual conditions relating to income distribution, which are deeply rooted in limited opportunities and constraints confronting individuals at the structural level. This inequality arises from geographical and familial circumstances at birth and during childhood, further complicated by the intersection of gender, race, location, and parental education. These differences create expected inequalities in income distribution (Sulla et al. 2022). The conditions of income inequality are evident in South Africa through the intergen-

M. Nathane (✉)
Department of Social Work, University of the Witwatersand, Johannesburg, South Africa
e-mail: Motlalepule.Nathane@wits.ac.za

N. V. Roman et al. (eds.), *Social Inequalities Within and Among Countries*, Sustainable Development Goals Series, https://doi.org/10.1007/978-3-032-08741-6_10

erational transmission of poverty and disadvantage. The bottom 50% of the population earns only 5–10% of the national income, particularly in contrast to the top 10%, which earns almost 85% of the national income in South Africa (Chancel et al. 2022).

Inequalities are evident along gender and racial lines due to the legacy of colonisation and apartheid. Households have been, and continue to be, adversely affected, particularly because many policies in the past discriminated against certain groups of families. The cruellest legacy of apartheid was how it worsened the conditions of Black African and Coloured children living in households with absolute poverty, leaving them without any form of state intervention (Streak 2010).

The lives of many women in South Africa are primarily characterised by raising children in households where biological fathers are non-resident (Datta 2007; Rabe 2006). According to the 2022 Census Report, South Africa's population stands at 62 million, comprising 17.8 million households (StatsSA 2022). The report further indicates that 49.6% of these 17.8 million households are classified as female-headed, while only 50% are headed by men. While it cannot be simplistically assumed that all female-headed households in South Africa are poor, the reality is that, in the peri-urban townships where this research was conducted, most of these households are indeed impoverished. South Africa's history of racial discrimination makes it essential to consider the diverse experiences of female-headed households.

10.2 Literature Review

The global trend regarding poverty currently indicates that 1 in every 10 women is living in extreme poverty (10.3%), with the majority—approximately 220.9 million—residing in sub-Saharan Africa (SSA) (United Nations Department of Economic and Social Affairs [UNDESA] 2023). Furthermore, more than one in every four women worldwide has experienced moderate or severe food insecurity. Without intervention to eradicate poverty, food insecurity rates in SSA could rise to 32.5%, leaving women particularly vulnerable by 2030. In this study, the term 'female-headed household' refers to households led by women at the time of the study. It also denotes a household where a woman, who is divorced, widowed, or has never married, is raising children under the age of 18 and is solely responsible for their care and economic provision (Wright et al. 2013; Payne and Range 1998). Even though women are not a homogeneous group and there are differences in class and economic status among them, evidence from around the world, particularly in SSA, shows that poverty is becoming increasingly feminised, with a growing proportion of the world's poor being female (Benería and Feldman 1992). This is further supported by the experiences of women who head households in the study conducted in Evaton. The poverty status of these women is evident not only in their immediate households but also in other households where they take on the role of head due to dependent children and a lack of additional income.

The South African government has made strides in mitigating high poverty levels by implementing various targeted state interventions, including direct cash transfers to families (World Bank 2018; Patel et al. 2023). Social assistance has been hailed as the largest state-targeted intervention in Africa. These cash transfers have had a significant impact on families in the developing world. For instance, Bolsa Família, a state-run conditional cash transfer (CCT) programme in Brazil, continues to contribute to the nation's efforts to improve income distribution (World Bank 2018). Social grants play a crucial role in reducing income poverty and hunger in poor households in South Africa. They have resulted in improved child health, nutrition, schooling, and the financial empowerment of caregivers. According to the South African Social Security Agency (SASSA), there is evidence that grants have also provided seed money to generate additional family income, make investments in education, and generally alleviate the stress individuals often experience due to the economic hardships associated with poverty (SASSA 2022).

10.2.1 The Feminisation of Poverty and Vulnerability in Households

South African women are vulnerable to poverty, a situation influenced by socio-historical structural factors and embedded patriarchal beliefs and values (Statistics South Africa [StatsSA] 2017). Women residing in rural and peri-urban townships are more exposed to poverty compared to their counterparts. Poverty and vulnerability are dominant features in female-headed households. Although poverty and vulnerability are two different concepts, they are inextricably linked, especially in the context of female-headed households (Chant 2008). Vulnerability, at a micro level, is a state of insecurity that the household and those who manage it experience as a result of being poor or when it seems they will remain in that state for an extended period. In contrast, poverty encompasses more than just the lack of income or material resources necessary for survival within a household. This is the lived reality for many women in South Africa who are raising children in environments with limited income opportunities. The government's investments in social grants and basic services have improved the lives of many South Africans (Patel et al. 2023).

10.3 Research Methods

This study employed a qualitative research method, commonly associated with the interpretive paradigm, as it follows a non-linear research path (Creswell and Creswell 2018). A narrative design was utilised to bring together various stories and actions from participants' lived experiences (Chase 2005). Purposive sampling employed, allowing the researcher to deliberately select participants in alignment with the research problem. The total sample consisted of thirty (30) participants. Household dyad interviews were conducted with fifteen (15) women and fifteen (15) young people from the same households. Face-to-face interviews were conducted individually, and both women and young people from each household provided separate informed consent to participate in the study. According to Fontana and Frey (2000), narrative interviews are less structured and aim to give participants the opportunity to share their stories without interruption. Narrative analysis was used to analyse the raw data.

10.3.1 Ethical Considerations and Research Site

Ethical clearance for the study was obtained from the University of the Witwatersrand's (Wits) Non-Medical Ethics Committee. Participants were informed that their participation was voluntary and that they had the right to withdraw from the study without facing any negative consequences. Full information about the purpose of the study and the methods used to obtain data was provided to all participants. Informed consent was obtained from mothers and from young people over the age of eighteen. Participants were also given consent forms to provide written consent for participation and for their interviews to be audio-recorded.

10.3.2 Research Site

Evaton was established in 1904 and celebrated its centenary in 2004, making it one of the oldest townships in South Africa. It is located 52 km south of Johannesburg and 24 km north of Vereeniging, a town that has remained a symbol of unity and peace since the end of the Anglo-Boer War. Evaton was founded during a time when South Africa was still divided into colonies, each with its own parliament that passed legislation (Department of Co-operative Governance and Traditional Affairs [CoGTA] 2009).

10.4 Findings

The findings indicate that poverty runs deep in these households, with the only stable source of income being social protection Child Support Grants (CSGs) and stipends earned by women through their voluntary work in local non-governmental organisation (NGOs). All of the women were unemployed, had limited income opportunities, and described their living conditions as a '*difficult life*'.

10.4.1 Difficult Life and the Burden of Provision

Poverty was a dominant feature across all households in this study. Participants consistently described a general theme of a '*difficult life*', particularly in relation to material provision within their households. The '*difficult life*' was characterised by vivid descriptions of the consequences of extreme poverty. Participants reported experiencing high levels of anxiety and stress, directly linked to persistent ill health, social isolation, malnourished children, overcrowded living conditions, and limited sustainable means of income generation.

The women narrated their experiences of a difficult life with deep emotion, often accompanied by visible pain, tears, and emotional breakdowns during interviews. The conditions of poverty and lack of income significantly affected the day-to-day provision of basic needs for household members. One participant, Fodisang, described 'provision' as a 'yoke', symbolising the burdensome nature of meeting the daily needs of family members, particularly children. Provision was perceived as a burden because it involved constant struggle and was characterised by ongoing hardship. Fodisang, a mother from household #2, was in her late 40 s at the time of the interview and had two children. Reflecting on her experience of raising her children through hardship, she said:

> "As a mother, my hardest challenge was raising my children whilst I tried my best to find employment. Finding employment was tough and I was mostly unemployed …, there were times when we went to bed on empty stomachs because there was no food to eat, no money, nothing at all. I was unemployed for a long time" (Fodisang).

Fodisang's narrative highlights the struggle of providing for children in the context of poverty and unemployment—an experience common in many South African households. In a context where wage employment is scarce for the majority of the South African population, the poor have increasingly retreated to their households, which have become their primary source of stability and survival. However, these households themselves are far from secure, facing overwhelming vulnerabilities related to food security and the lack of essential basic services (Mosoetsa 2011). Most families have either unemployed adults or only one employed person responsible for supporting many dependents. This exacerbates the challenges within these households, which tend to be larger in size, support more children, and suffer significant hardship when the sole employed adult loses their job. Furthermore, households facing challenges similar to Fodisang are often located in peri-urban townships and rural areas (May et al. 1998). The classification of life in poverty as a '*difficult life*' was also evident among participants in the 1997 South African Participatory Poverty Assessment (SA-PPA).

10.4.2 Pushing Life with CSG and NGO Stipend

Participants described their experiences as '*pushing life*', a phrase frequently used to characterise the realities of poverty and the struggles women face in providing for their families. This expression is also prevalent in public discourse, particularly in urban townships, to describe a life marked by difficulty and poverty. Consequently, the households in this study can be categorised as living below the national poverty line, with the CSG serving as their only stable source of income. The excerpt below illustrates similar dynamics of struggle through the experience of one participant, Anethe.

> "Our life is just a struggle; we push life with grant money because my mother is not employed. We stayed together there as a family pushing life with the grant money. Sometimes my mother would come to us just to see if we were still alive".

Raising children in the context of limited income opportunities and reliance on social grants was consistently described as 'pushing life'. In Anethe's words, *'the only way we are pushing life is through the Child Support Grant'*. In other words, the only stable income the household has is the CSG.

Most participants in this study were unemployed or underemployed, meaning they were available to work but unable to find employment. As a result, many participants volunteered at a local community-based organisation and received a monthly stipend of R1, 200. This stipend depended on the availability of funding from international donors and the Department of Health and Social Development. During periods when organisations ran out of funds, households faced hunger, as families primarily survived on state social grants. Therefore, CSGs became the only stable income for these households.

10.5 Social Intervention

In the context of high unemployment levels in South Africa, particularly among women in peri-urban townships like Evaton, this chapter proposes a social intervention that facilitates entrepreneurship and skills acquisition for women in the production of essential food items needed in communities. This initiative represents a form of self-help, where the government can partner with local private businesses to offer micro-lending programmes as seed funding for women to start small income-generating initiatives. There is a demand for locally produced food items in various communities at affordable prices, including bread, poultry products, and vegetables. The production and distribution of these items at the local community level will address food security, hunger, and high food prices. The unique challenge of food security in South Africa is that while food is widely available, it is not affordable.

10.6 Conclusion

It is evident from the findings discussed in this chapter that poverty, unemployment, and, in some cases, low-paid work, along with a lack of livelihood opportunities, continue to rob female-headed households of the capacity to provide for their dependents and force them to rely on state social assistance. This reliance leads to dependence on state social services, increasing the vulnerability of households and their inability to defend themselves against harsh economic conditions. High levels of the impoverished population in South Africa, particularly women, engage in work daily; however, this work is not necessarily classified as formal employment, and in some cases, they receive little pay for it. It is paramount that this position of vulnerability be reviewed alongside the structural issues surrounding the women who live in poverty, to avoid pathologising female heads of households as passive recipients of state social assistance. The women interviewed in this study were not helpless or inactive individuals dependent solely on social security, as is often portrayed in the media and wider society. Instead, these women engaged in voluntary work through various community-based organisations, providing care for the elderly, the sick, vulnerable children, and orphans at the local Early Childhood Development Centre. This finding challenges the prevailing binary view and definition of work, which categorises individuals as either employed or unemployed. In reality, much of the impoverished population in South Africa, particularly women, engage in daily work; however, this work is often not recognised as formal employment and, in many cases, is minimally compensated (Alexander and Wale 2013).

References

Alexander P, Wale K (2013) Underemployment: too poor to be unemployed. In: Alexander P, Ceruti C, Motseke K et al (eds) Class in Soweto. UKZN Press, Scottsville, pp 127–141

Benería L, Feldman S (1992) Unequal burden: economic crises, persistent poverty and women's work. Michigan: Westview Press

Chancel L, Piketty T, Saez E et al (2022) World inequality report 2022. World Inequality Lab, Paris

Chant S (2008) The "feminisation of poverty" and the "feminisation" of anti-poverty programmes: room for revision? J Dev Stud 44(2):165–197

Chase SE (2005) Narrative inquiry: Multiple lenses, approaches, voices. In: Denzin NK, Lincoln YS (eds) The sage handbook of qualitative research, 3rd edn. Sage, Thousand Oaks

Creswell JW, Creswell JD (2018) Research design: qualitative, quantitative, and mixed methods approaches. Sage, Thousand Oaks

Datta K (2007) In the eyes of a child, a father is everything: changing constructions of fatherhood in urban Botswana? Women Studies International Forum 30 (2), 97–113

Department of Co-operative Governance and Traditional Affairs (CoGTA) (2009) Township transformation timeline. CoGTA and the European Union, Pretoria

Fontana A, Frey J (2000) Qualitative methods: social science research. Sage, Thousand Oaks

May J, Woolard I, Klasen S (1998) The nature and measurement of poverty and inequality. In: May J (ed) Poverty and inequality in South Africa: meeting the challenge. David Phillips Publishers, Cape Town, pp 19–50

Mosoetsa S (2011) Eating from one pot: the dynamics of survival in poor households in South Africa. Wits Press, Johannesburg

Patel L, Dikoko V, Archer J (2023) Social grants, livelihoods and poverty responses of social grant beneficiaries in South Africa: research brief. Centre for Social Development in Africa, Johannesburg

Payne J, Range M (1998) Lone parents' lives: an analysis of partnership, fertility, and household composition in rural South Africa. Demography 49(2):699–718

Rabe M (2006) Being a father in a man's world: the experiences of gold mine workers. In: Richter L, Morrell R (eds) Baba: men and fatherhood in south. Africa. (Pages). HSRC Press, Cape Town

South African Social Security Agency (SASSA) (2022) SASSA 2020/21 annual report. SASSA, Pretoria

Statistics South Africa (StatsSA) (2017) Poverty trends in South Africa: an examination of absolute poverty between 2006 & 2015. Statistics South Africa, Pretoria

Statistics South Africa (StatsSA) (2022) Quarterly labour force survey (QLFS) Q2:2022. Statistics South Africa, Pretoria

Streak J (2010) Child poverty in South Africa and the performance of the child support grant programme. (Unpublished doctoral dissertation) Stellenbosch University

Sulla V, Zikhali P, Cuevas P (2022) Inequality in southern Africa: an assessment of the southern African customs union. World Bank Group, Washington, DC

UN Women, United Nations Department of Economic and Social Affairs (UNDESA) (2023) Progress on the sustainable development goals: the gender snapshot. UN Women and United Nations Department of Economic and Social Affairs, Statistics Division, Washington, DC

Wright G, Noble M, Ntshongwana P, Barnes H, Neves D (2013) Lone mothers in South Africa: the role of social security in respecting and protecting dignity. Johannesburg: Centre for the Analysis of South African Policy

World Bank (2018) An assessment of drivers, constraints and opportunities: overcoming poverty and inequality in South Africa. https://www.worldbank.org. Accessed 18 Mar 2025

Livelihood Precarity: Social Marginalisation and Resource Degradation Among the Seharia in Rajasthan, India

11

Bhavya Jain

Abstract

This chapter highlights the livelihood conditions of the Seharia, a Particularly Vulnerable Tribal Group in Rajasthan, India. The Seharia struggle due to inadequate resources, limited employment opportunities, and poor nutrition. Historically, they have depended on forests; however, this is changing due to changes in forest-related policies and composition. Given this situation, the study examined the livelihood-related problems faced by the Seharia in Rajasthan. The research employed a qualitative exploratory design, utilising in-depth interviews with community members as a method for data collection. This chapter is based on interviews with 13 community members. The main themes emerging from the findings include forest degradation, land grants and land grabs, and bonded labour, all of which profoundly affect the conditions of the Seharia community. The data indicate that the Seharia community experiences persistent socio-economic exclusion, making the achievement of the Sustainable Development Goal (SDG) of reducing inequality and promoting inclusion elusive in their context. In response, the study proposes the adoption of a producers' organisation or cooperative model as a potential strategy to enhance their current livelihood conditions.

Keywords

Sustainable livelihood · Seharia · Indigenous group · Precarity

11.1 Introduction

A stable livelihood is foundational to a good quality of life. However, many communities continue to face significant challenges in sustaining stable livelihood practices. Indigenous groups, including those in India, often have a heightened dependence on natural resources for their survival. Moreover, the intersection of their social identity as *Adivasi* (a term broadly corresponding to 'indigenous') and their limited access to property and wealth ownership increases their vulnerability, making them more reliant on precarious and less stable means of livelihood (International Labour Organisation [ILO] 2019; Li 2023). Although efforts have been made by the welfare state and other organisations to address their vulnerabilities, various socio-economic factors play a role in maintaining structures that sustain existing precarity and instability for these groups leading to persistent social inequalities and injustice.

The Adivasi groups in India have traditionally relied on forests and other natural resources for their livelihoods, but increasing regulation and resource degradation have forced many to

B. Jain (✉)
School of Social Work, Tata Institute of Social Sciences, Mumbai, Maharashtra, India
e-mail: bhavya181193@gmail.com

N. V. Roman et al. (eds.), *Social Inequalities Within and Among Countries*, Sustainable Development Goals Series, https://doi.org/10.1007/978-3-032-08741-6_11

seek alternative sources of income (Sundar 2016; Government of India 2013). Stable employment opportunities and sustainable livelihood practices are generally lacking for these communities (Jain 2019). This persistent precarity and resulting socio-economic exclusion may stem from the interplay between short-sighted policies that fail to translate into practice on the ground and long-standing social relations between Indigenous and other social groups. Therefore, working towards SDG 10.2 and SDG 10.3—i.e. ensuring the inclusion of all through equitable and just laws and policies—might require a closer examination of the intersectional realities of these groups and an enhancement of their participation in decision-making and execution.

This chapter draws on a study conducted with the Seharia community in India to examine the livelihood challenges and poverty-related struggles they face, particularly in the light of SDG 10: reduced inequalities within and among countries. The Seharia are the largest Particularly Vulnerable Tribal Group (PVTG)[1] in the country (Singh 2017), and the only PVTG residing in the North-Western province of Rajasthan. The Seharia community in Rajasthan is located exclusively in one administrative district—Baran. This chapter explores the factors behind the livelihood challenges faced by the community and suggests ways to address some of the inequalities they encounter. It begins by introducing the Sustainable Livelihood Framework, reviewing existing literature on the Seharia, followed by the methodology, key findings of the study, and intervention suggestions.

11.2 Literature Review

Historically, many PVTGs have engaged in the collection of forest produce and shifting agriculture as primary livelihood practices (Bhatia 2012). The Seharia, a historically nomadic community dependent on forest resources, traditionally did not 'own' much land. Their practice of shifting cultivation was discontinued with the introduction of settled and systematic land use and ownership systems (Tyagi and Padhi 2019). In the post-colonial period, the state attempted to address this resource deprivation through land grants; however, the land allocated was often of poor quality and not highly cultivable (Tyagi and Padhi 2019). Moreover, the Seharia's outcaste socio-historical position within a caste-based society rendered them vulnerable to systemic exploitation (Mahalwal and Kabra 2023). In Rajasthan's Kishanganj and Shahabad blocks alone, at least 982 cases of land alienation involving the Seharia have been documented (Philipose 2014). Today, the Seharia continue to depend on forests to meet a range of subsistence and livelihood needs.

Agricultural or Mahatma Gandhi National Rural Employment Guarantee Scheme (MGNREGS)[2] labour is a major source of cash income for the community (Soni 2014). The community has very low employment rates, with 66% of the Seharia involved in agricultural labour (Soni 2014). They have subsisted by adopting a variety of income-generating activities such as agricultural wage labour, migratory seasonal labour, debt bondage, and land cultivation (Patel and Mitra 2023; Samonova 2023). Although these activities have provided economic survival, they remain precarious, lacking sustainability for the community (Tyagi and Padhi 2019). Studies involving the same community in Madhya Pradesh have identified a significant loss of income and livelihood support over the past two decades, attributed to declining forest cover—particularly of firewood-yielding trees—and increasing restrictions on the access to and collection of Non-Timber Forest Produce (NTFP) (Mahalwal and Kabra 2023).

Livelihood precarity has had far-reaching consequences for the Seharia, affecting multiple

[1] According to the Ministry of Tribal Affairs, Government of India, a Particularly Vulnerable Tribal Group (PVTG) is defined as a community with a declining or stagnant population, low literacy levels, use of pre-agricultural technology, and pronounced economic disadvantage (Government of India 2023). As of now, India officially recognises 75 such PVTGs.

[2] Mahatma Gandhi National Rural Employment Guarantee Scheme (MGNREGS) is a flagship scheme of the Indian government, legally guaranteeing wage employment for hundred days in a financial year to employment-seeking adult members of a rural household.

aspects of their well-being. The community experiences high levels of malnutrition and disease burden (Patel and Mitra 2023). National attention was drawn to the Seharia in Rajasthan nearly two decades ago, following reports of several 'hunger deaths' (Jha 2002; Soni 2014). In response, a series of ad-hoc interventions were introduced. Existing research has primarily focused either on the health-related challenges faced by the community (Nigam et al. 2021; Bhat et al. 2022) or on their subsistence practices in Madhya Pradesh (Tyagi and Padhi 2019; Samonova 2023). This study extends the literature by examining the livelihoods and subsistence strategies of the Seharia through the lens of the Sustainable Livelihood framework and in relation to the SDG of addressing inequalities.

The literature reviewed above reflects the instability of livelihood practices and the heightened vulnerability experienced by the Seharia community. It also reveals the persistent discrimination they face and the inequalities evident in income distribution and access to opportunities. Building on these insights, this chapter argues that the Seharia continue to face significant barriers in accessing essential resources—barriers that are tooted in their current lack of material ownership and their historical status as a forest-dwelling and dependent community.

11.3 Conceptual Framework

The study adopted the Sustainable Livelihood Framework, developed by Chambers and Conway (1992), to examine the livelihood practices and poverty experienced by the Seharia community. According to their framework,

> Livelihoods comprise of capabilities, activities and assets required for a means of living. Sustainable livelihood can cope with and recover from stress and shocks, maintain or enhance its capabilities and assets, and provide sustainable livelihood opportunities for the next generation; and which contributes net benefit to other livelihoods at the local and global levels and in the short and long run. (1992: 6)

Thus, any livelihood practice involves people and the activities they undertake using their knowledge, skills, and creativity to mobilise the assets they possess in order to produce essential outcomes such as food, income, and other necessities. This output may be further improved through the reinvestment of surplus. However, for a livelihood to be considered sustainable, three fundamental elements must be present:

- *Capability*—This refers to the ability of individuals involved in livelihood activities to cope with shocks and stresses, and to pursue new and improved opportunities through innovation, collaboration, and adaptive strategies.
- *Equity*—This refers to the fair distribution of assets, opportunities, and capabilities among those involved in a livelihood activity, with particular attention to the needs of the deprived, ensuring that all participants are able to sustain themselves within the livelihood system.
- *Sustainability*—This refers to maintaining and improving livelihood practices and assets, both tangible (physical and material such as land, water, and money) and intangible (access and claims).

Another fundamental element that may be added, as suggested by Krishna (2012), is *collective political action*. This component is particularly significant for marginalised populations, as collective action strengthens their ability to access and exercise control over vital resources. For indigenous communities, the lack of access and voice in matters concerning resource loss or degradation is exacerbated by the intersection of their indigenous identity and socio-economic class.

The absence of these fundamental elements underscores the precarity of a livelihood practice, reflecting limited access to resources, the degradation of existing assets, and broader patterns of marginalisation. *Precarity* refers to heightened levels of uncertainty, insecurity, and instability in economic activities (Maiti 2012), often experienced by groups whose historical social status has shaped their current, disadvantaged position within the labour market (Mallett 2020). Precarious work situations are characterised by limited social protections and an inability to plan for a stable future (Walsh 2019). Informal, temporary, or contingent employment—forms of work that dominate the livelihood practices of the Seharia—are prime examples of such precarity (Kasmir and Stasch 2018). These precarious conditions have deep socio-historical roots, particularly in the Global South (Kasmir and Stasch 2018). This chapter examines how these dynamics manifest in the livelihood practices of Seharia in Rajasthan.

11.4 Methodology

This chapter draws on a qualitative exploratory study conducted during 2018–2019, based primarily on first-hand data collection. Participants were accessed through a local organisation engaged with the Seharia community. A combination of purposive, opportunistic, and criterion-based sampling strategies was employed to identify and select participants.

The data were collected through in-depth interviews using an interview guide. The guide included questions relating to livelihood practices, income sources, government schemes, and changes in daily routines. The sample comprised four men and nine women from the Seharia community. To capture regional diversity, participants were selected from various sub-regions within the blocks. The researcher resided in one of the villages for a period of four months to facilitate rapport-building and enhance the credibility and trustworthiness of the data collected.

The data were coded using Atlas.ti software, with emergent codes subsequently organised into sub-themes and overarching themes. The ethical principles of informed consent, self-determination, acceptance and no harm were diligently followed throughout the study. Participant names have been changed to protect their identities and maintain confidentiality.

11.5 Findings

The study found that the Seharia continue to struggle to find sustainable sources of livelihood. Although they have attempted to diversify their livelihood practices to sustain themselves, they remain trapped in a cycle of poverty, marginalisation, and exploitation. Their historical status as forest dwellers, combined with limited ownership of conventional forms of wealth and resources, has subjected them to systemic discrimination and contributed to a significant lack of opportunities. The major themes from the analysis presented here are forest degradation and engagement in bonded labour. This section discusses their experiences regarding various means of subsistence, highlighting a lack of capability, equity, sustainability, and collectivisation in their livelihood practices, which ultimately leads to precarity. It also underscores the role of structures such as social class, the modern welfare state, and hierarchy based on established social castes.

11.5.1 Forest Degradation

Forests have been a source of income, nutrition, and other essential needs for the Seharia community. However, this reliance is diminishing. There are two reasons why the Seharia can no longer depend on the forest resources as they once did. First, there has been a decrease in accessible forest cover. Second, while the overall forest cover has increased due to plantation drives from the Forest Department, the composition has changed, limiting the resources available to the community. This is illustrated in the following statement by Babita:

> ...Earlier people used to roam around more in the forest, so we brought vegetables from there...now the vegetables are not as good.... wood to make things such as a cot or door would come from the forest and now... has to be bought from the market. And then people would bring these leaves of tendu (Diospyros melanoxylon) to roll beedi (rolled Indian cigarettes), and achar, all these were the benefits of the forest.

Here, we observe a decrease in people's capabilities and the sustainability of a vital livelihood resource for the community. Sustainability is hindered by the changing composition of the forest. The Forest Department's control over the forests has resulted in the Seharia having a reduced claim to and access to these resources. Furthermore, the forest would offer them resilience to face any shocks by providing food and income.

However, decreased diversity and limited access to the forest have adversely affected the community's ability to meet its needs during times of distress.

Thus, the state's approach to forests, through the forest department, focuses on transforming them into commercially viable spaces, overlooking the local community's relationship with and dependence on these resources. While forest degradation affects everyone, the greater dependence of indigenous communities, such as the Seharia, places undue pressure on their survival strategies.

11.5.2 Hali—Labouring in Bondage

With degrading forests, inadequate land holdings, rising needs, poverty, and a lack of sufficient work, the Seharia are often indebted. Loans are common, but formal sources are not frequently accessed. The absence of a fixed income or capital diminishes the chances of approval for a formal loan. Moreover, income uncertainty makes timely loan repayment difficult, thereby creating a cycle of unsafe debts for the poor within the current financial system. This situation has forced the community into the practice of agricultural debt bondage (Samonova 2023). The system is colloquially known as '*Hali*'. Halis work for their debtors for 12 months or longer if the loan amount is higher. They work at a low wage rate of INR 100 per day (the regular wage is otherwise INR 150–250 per day). They are engaged in any activity and at any time of the day as required by the employer. Interest may also be levied on the lent amount. One of the participants, Shyama, elaborated,

> If one works as a hali, then the wage paid is hundred rupees per day. Now suppose at the time of settlement, if someone comes out as a debt on me, then they pay me only a hundred. But during the soybean harvesting season, I may get two hundred or 150 rupees as a wage per day.

The Bonded Labour (Abolition) Act of 1976 bans this practice in India; however, it continues nonetheless, reflecting ignorance and inefficiency on the part of the justice system. Engaging in *hali* usually involves the entire family. Such labour relations adversely affect all the fundamentals of a sustainable livelihood. They are exploitative and foster inequality by creating a strict hierarchy between the master and the *hali* (worker/slave).

11.6 Discussion

The previous section illustrates how the livelihood practices of Seharia fall short of the fundamentals of sustainable livelihoods—capability, equity, sustainability, and collective action. This absence of fundamentals results in precarity, as their sources of income are uncertain, with no social protection or better prospects for the future. (Walsh 2019; Maiti 2012). They continue to live in poverty and engage in practices such as *hali*, which violate decent work principles, as noted by Samonova (2023). This situation is both a cause and a consequence of the social inequality they face, highlighting the interconnectedness of the SDGs. While many Seharia exercise agency on a day-to-day basis, their historical socio-economic status obstructs their ability to voice opinions and coerces them to endure existing inequalities and injustices within the larger socio-political context (Jain 2019).

The degradation of surrounding forests has impacted the Seharia's survival, much like other forest dependent communities, and infringes upon their right to life and livelihood, as argued by Banerjee and Madhurima (2013). Moreover, the lack of long-term programmes aimed at rehabilitating those engaged in the *hali* system makes it difficult for them to find stable alternatives and break free from bondage. As Awasthi (2014) demonstrates in the context of Nepal, banning bonded labour must be accompanied by the provision of secure livelihood options to ensure that those who are rescued have stable sources of income and do not fall back into cycles of bondage. The Indian welfare state provides services and entitlements at both the central and provincial levels to the Seharia. These include free schooling and residential schools, services to address malnutrition, subsidised food rations, and reservations in government jobs. While these services have benefited the community, they fail to bring about structural change. As Yadav (2020) observes, a secure livelihood entails not only economic stability but also the enhancement of one's social position. However, services intended for the community are often short-sighted and populist, coupled with a general lack of transparency and accountability (Jain 2019). These challenges are further compounded by persistent social discrimination and prejudice (Jain 2019). As a result, the Seharia are able to achieve only minimal economic sustenance, without any real sense of security or social advancement.

One limitation of the study is that it does not examine the functioning and success of certain farmer-specific programmes run by the government. Additionally, the study participants were accessed through the host non-governmental organisation (NGO), which may have introduced a degree of selection bias. NGO staff directed me to individuals they knew personally or had worked with through the NGO for several years.

11.7 Proposed Intervention

The problems faced by Seharia regarding their livelihoods relate to a lack of adequate resources, socio-economic marginalisation, precarious alternative livelihood practices, and insufficient collectivisation.

The formation of collectives could serve as an effective intervention to address these issues. Collectives have been established for agricultural produce, forest products, dairy, and other industries in India (Harrington et al. 2023). A similar collective or producers' cooperative could be established with Seharia to enhance their income, provide affordable loans, and empower them socially. The organisation initially focuses on identifying common needs, skill development, and knowledge building, while also establishing best practices and planning activities. It identifies potential markets/customers and establishes important linkages. Collectivisation can open new possibilities for bargaining with other parties. Sustained engagement may also lead to enhanced decision-making capabilities for individuals, especially for women, who are often excluded from livelihood-related decision-making in the community (Jain 2019). As the organisation grows, it can work towards larger social goals, as has been the case with some well-established organisations (Harrington et al. 2023), and serve as a platform to voice the opinions and experiences of discrimination that the community wishes to highlight.

Any such collectivisation would require support at its inception to bring people together and establish the foundational structures and values of the organisation. Several challenges are likely to arise at the outset, including issues related to infrastructure, prevailing gender and social norms, and initial apprehension among potential members (Gupta 2016).

11.8 Conclusion

A decent livelihood is a fundamental right for every individual and community, as it is intrinsically linked to overall quality of life. The absence of stable livelihood opportunities can intensify existing forms of discrimination and further restrict access to resources and opportunities, particularly in poverty-stricken contexts. Therefore, it is imperative that efforts be undertaken in close collaboration with the Seharia community to enable access to their traditional resources and ensure livelihoods that are free from exploitative conditions and provide adequate income to alleviate poverty. A context-specific intervention based in community collectivisation and aiming for economic and social empowerment is needed to address the deep rooted social inequality and injustice faced by the Seharia. In this regard, the establishment of a producers' organisation or cooperative offers a viable pathway for achieving these goals and advancing the SDGs related to ending poverty, promoting decent work, and reducing inequalities within the community.

References

Awasthi N (2014) Freed bonded labour in the far Western Tarai. Third Pole J Geogr Educ 11:29–33. https://doi.org/10.3126/ttp.v11i0.11527

Banerjee A, Madhurima C (2013) Forest degradation and livelihood of local communities in India: a human rights approach. J Hortic for 5(8):122–129. https://doi.org/10.5897/JHF2013.0305

Bhat J, Yadav R, Sharma RK et al (2022) High incidence of pulmonary tuberculosis in an indigenous Saharia tribe in Madhya Pradesh, Central India—a prospective cohort study. PLOS Glob Public Health 2(6):e0000039. https://doi.org/10.1371/journal.pgph.0000039

Bhatia B (2012) Of human bondage in Baran, Rajasthan. Econ Polit Wkly 47(26):159–169

Chambers R, Conway G (1992) Sustainable rural livelihoods: practical concepts for the 21st century. IDS Discuss Paper 296. Institute of Development Studies, Brighton

Government of India (2013) Statistical profile of Scheduled Tribes in India. Ministry of Tribal Affairs, New Delhi

Government of India (2023) Annual report 2022–23. Ministry of Tribal Affairs, New Delhi

Gupta H (2016) Farmer producer organization: enhancing the identity of women as farmers. https://www.pradan.net/sampark/farmer-producer-organization-enhancing-the-identity-of-women-as-farmers. Accessed 18 Mar 2025

Harrington T, Narain N, Rao N et al (2023) A needs-based approach to promoting gender equity and inclusivity: insights from participatory research with farmer-producer organisations (FPOs). J Soc Econ Dev 26:409–434. https://doi.org/10.1007/s40847-023-00280-x

International Labour Organisation (ILO) (2019) Decent work for Indigenous and Tribal Peoples in the rural economy. https://primarysources.brillonline.com/browse/climate-change-and-law-collection/decent-work-for-indigenous-and-tribal-peoples-in-the-rural-economy;cccc019620190196524. Accessed 18 Mar 2025

Jain B (2019) Changing livelihood practices among Sahariyas in Rajasthan. Tata Institute of Social Science, Mumbai

Jha MK (2002) Hunger and starvation deaths: call for public action. Econ Polit Wkly 37(52):5159–5163

Kasmir S, Stasch R (2018) Precarity. In: Stein F, Lazar S, Candea M, Diemberger H, Robbins J, Sanchez A, Stasch R (eds) Cambridge encyclopedia of anthropology. Department of Social Anthropology, University of Cambridge, Cambridge

Krishna S (2012) Redefining sustainable livelihoods. In: Harcourt W (ed) Women reclaiming sustainable livelihoods: spaces lost, spaces gained. Gender, development and social change series. Palgrave Macmillan UK, London, pp 12–18. https://doi.org/10.1057/9781137022349_2

Li T (2023) Precarious livelihoods and the challenges of collective action. https://sites.tufts.edu/reinventingpeace/2023/11/08/precarious-livelihoods-and-the-challenges-of-collective-action/. Accessed 18 March 2025

Mahalwal S, Kabra A (2023) The slow violence of fortress conservation creates conditions for socially unjust "voluntary" relocation. Biol Conserv 286:110264. https://doi.org/10.1016/j.biocon.2023.110264

Maiti D (2012) Precarious work in India: trends and emerging issues. Am Behav Sci 57(4):507–530. https://doi.org/10.1177/0002764212466247

Mallett RW (2020) Seeing the "changing nature of work" through a precarity lens. Glob Labour J 11(3). https://doi.org/10.15173/glj.v11i3.4138

Nigam S, Sharma RK, Yadav R et al (2021) Experiences and needs of patients with MDR/XDR-TB: a qualitative study among Saharia tribe in Madhya Pradesh, Central India. BMJ Open 11(8):e044698. https://doi.org/10.1136/bmjopen-2020-044698

Patel PK, Mitra RP (2023) Prevalence of chronic food insecurity, policies, and redressal patterns among the PVTGs: a case study of Sahariya tribe from India. Reprint at https://doi.org/10.21203/rs.3.rs-2620991/v1

Philipose P (2014) Tomorrow will not be like yesterday': Sahariya tribals emerge from the shadows. https://eldis.org/document/A100696. Accessed 18 March 2025

Samonova E (2023) Lifeworlds of female bonded labourers among the Sahariya tribe. Dev Pract 33(2):205–214. https://doi.org/10.1080/09614524.2022.2153802

Singh SS (2017) Vulnerable tribes: lost in a classification trap. https://www.thehindu.com/news/national/vulnerable-tribes-lost-in-a-classification-trap/article17894997.ece. Accessed 18 March 2025

Soni V (2014) A study on hunger deaths among Saharia of Baran, Rajasthan. Tata Institute of Social Science, Mumbai

Sundar N (2016) Introduction: of the scheduled tribes, states, and sociology. In: Sundar N (ed) The scheduled tribes and their India: politics, identities, policies and work. Oxford University Press, New Delhi, pp 1–46

Tyagi M, Padhi SR (2019) Livelihood diversification and challenges among Sahariya tribe in Gwalior District. Think India J 22(14):2251–2261

Walsh P (2019) Precarity. ELT J 73(4):459–462. https://doi.org/10.1093/elt/ccz029

Yadav S (2020) Precarity as a coping strategy of the Gonds: a study of insecure and long-distance seasonal migration in Central India. Indian J Hum Dev 14(1):7–22. https://doi.org/10.1177/0973703020909328

Unfolding Gond Women's Struggles and Agency in the Traditional Livelihood of the Bamboo Craft Practice

12

Seema Panchabhai

Abstract

The Gond tribe is one of the marginalised and vulnerable groups within the hierarchy of the Indian social system. Historically, Gond women have demonstrated resilience and agency, actively contributing to their everyday lifeworlds. Gender relations within the community are shaped by daily practices, which can be both empowering and oppressive for women. Among traditional livelihood activities, bamboo craft is a seasonal practice in which women play a significant role. However, the significance of Gond women's labour often remains unrecognised, leading to their contributions being undervalued and their status within families and communities largely unacknowledged. This study seeks to reveal the full extent of Gond women's contributions to women's agency and the interface of space within bamboo craft activities. A phenomenological approach was employed to understand the lived experiences of Gond women from three villages located in the Korchi block of Gadchiroli district, India. Using an intersectionality approach, this chapter foregrounds their struggles and delineates their agency in transforming traditional livelihood structures. Aligned with Sustainable Development Goal (SDG) 10--reduced inequalities—this study emphasises the importance of social interventions at the community level. By drawing on community values and harnessing traditional skills and knowledge, Gond women can be empowered, thereby opening new livelihood opportunities. Community-based interventions play a critical role in cultivating values of gender equality at both family and community levels.

Keywords

Agency · Bamboo craft · Gender · Sustainable livelihood · Tribal

12.1 Introduction

Globally, tribal communities remain among the most deprived and oppressed population groups compared to other social categories (Dasgupta 2018; Xaxa 2016). In India, the Gond tribe is one of the largest tribal groups, traditionally engaged in agriculture and subsistence activities such as forest produce collection, fishing, and bamboo craft—occupations in which women participate actively and equally. Tribal women play a central role in the economic life of their communities (Shaik 2016). However, from an intersectionality perspective, tribal women hold a secondary status within the community. Bamboo craft is one of the seasonal livelihoods that creates a supplementary

S. Panchabhai (✉)
School of Social Sciences, Tata Institute of Social Sciences, Mumbai, Maharashtra, India
e-mail: seema.tissphd2019@gmail.com; dr.simapanchabhai@gmail.com

N. V. Roman et al. (eds.), *Social Inequalities Within and Among Countries*, Sustainable Development Goals Series, https://doi.org/10.1007/978-3-032-08741-6_12

income source, enabling tribal people to survive. At the same time, tribal women have struggled to gain ownership of natural resources, which has led to survival and social inequality (Krishna 2001).

The Gadchiroli district of Maharashtra is a forest and tribal area, where the Gond community mainly engages in agricultural activities. In the Korchi block, both Gond men and women participate equally in seasonal bamboo craft activities; however, women's contributions are frequently overlooked. A phenomenological approach is employed to interpret the community's daily lifeworld experiences, which inform their values related to equality. This approach enables a critical examination of Gond women's agency and individual perspectives. The chapter explores the challenges faced by women and provides a critical analysis of the market structures surrounding bamboo craft activities. This chapter reveals the agency of Gond women as they engage equally in these traditional livelihood activities. The Gond community structure and daily lifeworld are analysed based on primary data collection and participatory methods. This issue relates to Sustainable Development Goal (SDG) 10.3, which aims to ensure equal opportunity and end discrimination, as Gond women face discrimination based on gender and are not recognised for their traditional knowledge and skills within the community.

This study explores the struggles faced by Gond women in relation to bamboo craft and examines their agency within the community. The paper is structured into four sections. The first section outlines the relevant literature, methodological approach, and theoretical framework guiding the study. The second presents the key findings. The third section analyses the relationship between SDG 10 and the socio-economic status of Gond women, with particular attention to their involvement in bamboo craft activities. The final section offers social intervention strategies informed by the study's findings and concludes with key insights.

12.2 Literature Review

Reducing inequality lies at the heart of socially inclusive economic growth and is essential for the sustainable development of local communities. At the international level, SDG 10 aims to promote equal opportunities and eliminate discrimination, thereby advancing intergenerational equity (Biberman and Bajpai 2020). In India, there is a misconception that tribal women enjoy a higher status in their communities compared to non-tribal women. In reality, the majority of tribal women face inequality and discrimination in patriarchal tribal areas. The main reasons for this are the rigid culture, ritual servitude, and strict customs of their communities, which keep women suppressed and unaware of their rights (Khan and Hasan 2020). In contrast, Sumi Krishna explained that the tribal gender relationship is more egalitarian than that of the non-tribal community (Krishna 2001). Additionally, Krishna sought to explore the relationship between tribal women and natural resources. The dynamics of gender and ownership of natural resources have been evolving under the influence of the non-tribal community and changing policy structures (Krishna 2003). Baral's (2014) research on bamboo craft and tribal communities highlights the intense engagement of tribal women in traditional bamboo management and the creation and development of products. His study points to the potential for empowering these women through capacity building and training (Baral 2014). Similarly, Bain notes that in the Indian context, bamboo craft functions as a traditional livelihood for the Lepcha tribal community, contributing to the preservation of traditional knowledge and skills (Bain 2022). To frame these dynamics analytically, Crenshaw's (1992) concept of intersectionality offers a valuable lens for understanding black feminist issues through multiple lenses. At both academic and policy levels, the intersectionality perspective is understood as the convergence of social differences and power structures with the lived experiences of individuals (Sengupta and Harris 2022). In feminist studies, intersectionality is employed as a metaphor to articulate oppression through an open-ended approach (Carastathis 2014). This chapter applies Crenshaw's (1992) conceptualisation of intersectionality through three interrelated dimensions: gender, tribal identity, and livelihood. Gond women experience layered forms of marginality as they navigate the intersecting realities of being women, members of a tribal community, and inhabitants of forest-dependent regions. To understand their agency, Kabeer (1999) offers a nuanced view, suggesting that agency involves more than decision-making power; it includes the ability to bargain, negotiate, subvert, and resist, sometimes through strategies such as deception or manipulation. Agency, in this sense, is a cognitive and reflective process. In the context of Gond women's struggles, their agency—particularly in relation to bamboo craft livelihood activities—can be understood through Crenshaw's intersectionality framework, alongside the insights of Krishna (2001), Baral (2014), and Bain (2022) concerning natural resource ownership and socio-economic agency.

12.3 Methodology

12.3.1 Theoretical Framework

This chapter draws on three interrelated conceptual frameworks—intersectionality, agency, and sustainable livelihoods—to inform its methodological approach and analysis (Fig. 12.1; Crenshaw, 1992).

Crenshaw (1992) introduces *intersectionality* as a powerful lens for understanding how multiple forms of inequality and disadvantage intersect to create compounded forms of marginalisation. She explains: "Intersectionality is a metaphor for understanding how multiple forms of inequality or disadvantage sometimes compound themselves and create obstacles that often are not understood among conventional ways of thinking".

Building on this, Kabeer (1999) conceptualises *agency* as the ability of individuals to make purposeful choices and act upon them. She writes: "The ability to define one's goal and act upon them. Agency is more than observable action; it encompasses the meaning, motivation and purpose individuals bring to their activity, their sense of agency, or 'the power within'".

In relation to livelihoods, Chambers and Conway (1991) define *sustainable livelihood* as: "[T]he capabilities, assets and activities required for a means of living; a livelihood is sustainable and can cope with and recover from stress and shocks, maintain or enhance its capabilities and assets, and provide sustainable livelihood opportunities for the next generation; and which contribute net benefits to others livelihoods at the local and global levels and in the short and long term".

These three conceptual anchors are deeply interconnected. Together, they inform both the study's analytical lens and its methodological orientation, which is explained in Fig. 12.1.

This research was conducted using a qualitative research methodology and a phenomenological approach. The study population comprises Gond tribal people engaged in bamboo craft activities, with a specific focus on Gond tribal women. Geographically, the Korchi block is situated in the northern part of the Gadchiroli district in Maharashtra state, known as central India. The Gond community primarily resides in this area, living in small villages called 'padas' and following a joint family structure. The study was conducted in three villages in the northern region of the Korchi block in the Gadchiroli district of India.

A purposive sampling method with a saturation sampling technique was employed, involving communication with a resource person who had already participated in bamboo craft activities. From the selected pool of older and younger women, those willing to replace their bamboo

Fig. 12.1 Theoretical framework. *Note* Adapted from Crenshaw (1992)

craft activities with education were chosen. Additionally, young women actively involved in bamboo craft activities were also selected for the study. All participants were between 18 and 60 years of age.

The primary data collection methods included in-depth interviews, case studies, oral history, observation, and the Participatory Rural Appraisal (PRA) method. Within the PRA framework, transect walks, seasonal calendars, and focus group discussions were utilised to gather information on women's participation in various livelihood activities, as the Gond community engages in seasonal livelihood practices. Ethical considerations such as consent, confidentiality, and respect were upheld throughout the data collection process.

The study's reliability was assessed based on truth value, consistency, and neutrality, with the researcher aiming to understand women's agency from a tribal perspective. Primary data were transcribed and translated based on thematic analysis, from which codes, subcodes, themes, and subthemes were derived. Primary data analysis was conducted using Atlas.ti software.

12.4 Major Findings

Based on the primary data collection, the major findings explain that Gond women play a significant role in the bamboo craft-making process. However, Gond women struggle to sustain their agency in the seasonal livelihood activity. Markt is a significant structure in bamboo craft livelihood activity, where women have to find their own way to empower themselves at the family and community levels.

12.4.1 Gond Women's Involvement in the Bamboo Craft

Bamboo is a primary natural resource for the tribe and is used in their daily lives for various purposes, such as making bamboo baskets, kitchen sets, agricultural tools, fishing instruments, and hut construction. The Gond community relies on bamboo for their everyday needs and cuts and sells it to earn money for survival. In the past, women were actively involved in the bamboo-cutting process. However, contemporary deforestation and changing forest governance laws have decreased women's participation, as the village local governance body now strictly monitors and controls the forest area. Gond women are no longer allowed to access the forest as freely as they once did. Consequently, the new laws have limited women's agency and their traditional access to forest resources for survival. As a result, Gond men have taken on the responsibility of gathering bamboo from the forest, which has affected women's traditional knowledge regarding bamboo selection and its use for various crafts. This knowledge about the variety, type, and kinds of bamboo required for different bamboo crafts can no longer be transferred to the next generation of young Gond women.

In the Korchi block, bamboo craft is a traditional seasonal livelihood activity that helps sustain families during lean periods. Most Gond villagers engage in bamboo craft activities, which also help preserve their traditional knowledge. The women of the tribe equally participate in bamboo craft activities such as cutting bamboo, making canes, cleaning bamboo and canes, weaving bamboo products, preserving them from pests, and selling the final products. Despite the challenges faced by Gond women in accessing the forest, they attempt to gather bamboo with men at the beginning of the craft season. Gond women learn bamboo craft primarily through observation and the 'doing and learning' process, which are two methods used to acquire knowledge and skills. Women who move to the village after marriage also learn bamboo crafts and share equal responsibility in livelihood activities.

> I learned bamboo craft after my marriage. Now, I can weave various bamboo crafts. My husband sells this craft and manages this money for daily expenses. (Gond woman, 34-year-old)

During the bamboo craft season, women and men wake up at 3:00 a.m. Women complete household chores while men prepare the raw materials for making crafts until 9:00 a.m. From morning to evening, women weave bamboo products while men engage in agricultural activities. Primarily, women are involved in the weaving process after preparing the canes. Finally, men are responsible for completing each craft to make it ready for sale. Once they finish the weaving, women attend to the remaining household activities, while men prepare for the next day's bamboo craft products. The Gond women use traditional methods to protect and manage bamboo products from pests. They create various bamboo crafts, such as baskets of different sizes and shapes, fishing products, and agricultural items. Since it is the Gond men who sell the products, the women are unaware of the financial transactions and do not learn about them. Consequently, control over the money lies with the men of the Gond households. The Gond community overlooks the women's agency in economic decisions, despite their active participation in the craft-making process.

12.4.2 Gond Women's Struggles in the Bamboo Craft

In the bamboo craft activity, Gond women primarily focus on the weaving process rather than on collecting bamboo from the forest, cleaning it, and creating canes of various sizes. They produce a range of bamboo crafts using sharp knives. However, crafting canes of different sizes can be challenging, as it requires handling sharp knives to create both small and long canes. This fear of using sharp knives leads the women to rely on male members for assistance in making bamboo crafts.

> I have a fear of using sharp knives. Using a knife and making canes to make the bamboo craft is difficult. (Gond woman, 35-year-old)

Most Gond women are responsible for creating winnowers, which are in high demand during the specific bamboo craft season. These women produce winnowers in various sizes and shapes; however, most lack the necessary skills to give them the final touch and achieve the proper shape. Only a few women have learned this process through practical experience. Refining the final form and adding the finishing touch is a highly skilled endeavour. Although women typically acquire the art of bamboo crafting after marriage, they often face difficulties in mastering this skill. Conversely, those who begin learning bamboo crafting at a young age are able to refine their skills and produce a variety of items in different sizes, such as baskets and agricultural tools.

> Women make only half the winnower; then men complete it by giving shape to the corner, tying all canes, etc. Women do not know how to tie and give the shape. Some women tie the canes and give shape also. Not every woman can do this work. (Gond woman, 55-year-old)

However, industrialisation and the use of plastic, iron, and steel have influenced the production of bamboo products. Bamboo product-making activities are time-consuming and laborious and generate less income. Therefore, only some Gond is involved in bamboo craft activities. With this circumstance, Gond women manage their family structure based on natural resources.

12.4.3 Generational Shift: Youth Girls' Agency and the Bamboo Craft

The younger generation of the Gond community prefers wage work and finds traditional livelihood activities, such as bamboo crafting, difficult and unprofitable. Another reason is that young Gond girls are often away from home for formal education, which means they cannot observe their parents' activities. As a result, the informal 'self-learning' method of passing on traditional knowledge is diminishing, and the formal education process makes it challenging for them to maintain interest in adopting their cultural practices.

The young generation adopts bamboo craft skills and engages in it only to overcome financial difficulties. Young Gond girls do not participate in bamboo craft activities because, after marriage, they move to their in-laws' homes and do not take on the family's economic responsibilities. A young girl cannot make bamboo crafts if her in-laws' family is not involved in bamboo craft activities. Conversely, Gond young girls are not inclined to learn traditional livelihood activities as they are attracted to new technical skills and knowledge. Consequently, families involved in bamboo craft activities do not encourage girls to learn. The patriarchal system thus hinders the transfer of traditional knowledge to the next generation.

> Why do girls need to learn bamboo craft activities? After marriage, they will go to another village, so they do not require it. (A mother of a daughter)

Bamboo craft is a traditional livelihood activity that generates limited income, leading parents to be reluctant to involve the next generation in such pursuits. Consequently, young girls are more inclined to acquire new technical skills through the formal education system rather than through traditional activities. This trend represents a generational shift in the bamboo craft sector. Gond young girls form their opinions based on formal education and seek alternative ways to thrive in the modern world.

12.4.4 Restrictions on Gond Women's Role in Marketing

In the past, Gond women would go to the market with their family members to sell bamboo products. They carried their products on their heads and walked long distances of 20 to 25 km from the forest area to reach the market, as no means of transportation were available. However, intermediaries have become an essential part of the selling process. Economically prosperous families create bamboo crafts in bulk and sell them to middlemen at the village level. Occasionally, Gond men also sell their products in the market according to the prevailing market price. As a result, Gond villagers receive low prices from the middlemen for their efforts, skills, and knowledge. Typically, Gond men take the initiative to sell bamboo crafts to manage their household expenses. In the absence of men, women sell bamboo products according to the instructions given by family members. Nevertheless, women are skilled negotiators and often secure a higher price for bamboo craft products than men.

> Mostly, men sell the product; when women get this responsibility, they fight with middlemen to get a higher rate of bamboo products. (Gond woman, 35-year-old)

Despite playing a crucial role in bamboo craft activities, Gond women rarely make economic decisions.

12.5 Discussion

SDG 10 aims to reduce inequality and create equal opportunities through individual and community-based interventions (Biberman and Bajpai 2020; Oestreich 2018). This chapter centres on Gond women who practice bamboo crafts. Despite encountering gender inequality as women in society, these individuals also face unique challenges as members of a tribal community that relies heavily on natural resources for survival. Kabeer et al. (2023) have explained that Gond women have been overlooked in terms of their agency and potential. Their voices are rarely recognised within the community. The involvement of Gond women in bamboo craft activities is often disregarded, despite their participation in cutting, cleaning, and cane making. Nevertheless, Gond women have learned to adapt and engage in craft-making to ensure their survival, despite the limitations placed on their involvement.

Typically, women begin learning the bamboo craft after marriage and enjoy weaving, but they struggle with using sharp knives and completing bamboo products. However, more experienced Gond women possess the necessary skills to teach the younger generation through practical application. Kabeer (1999) explained that agency defines and acts upon its goals. Young Gond girls often prioritise bamboo craft as a secondary option and gravitate towards the formal education system to create better livelihood opportunities. Bayraktaroğlu (2020) identified the selection of the main actor within the community and highlighted innovative ways to lead to a rural craft economy, thereby opening new space for decision-making at both the family and community levels. Gond women are influential participants in the bamboo craft industry, significantly impacting the region's economy. They navigate the block-level market and middleman networks to make informed decisions and alleviate economic hardships at the family level.

12.6 Proposed Social Intervention

Based on the findings of this study, a targeted social intervention is recommended to promote gender equality and create sustainable livelihood opportunities for Gond women, who, despite their active participation in bamboo craft activities, continue to be relegated to a secondary status within their communities.

The Gond community has a long-standing practice of cooperation within their community. This collaborative approach has led to the establishment of Self-Help Groups (SHGs) at the village level, with the assistance of NGOs and government agencies. SHGs consist of young and older individuals (from 18 to 50 years old), with the intervention implemented in three phases: the first phase focuses on raw materials and skill development; the second phase involves the integration of traditional and technological knowledge, as well as the development of innovative products; and the final phase centres on marketing and ensuring the sustainable development of bamboo crafts.

The initial phase focuses on natural resources and enhances individuals' skills, knowledge, and techniques. The village Gram Sabha allows for a five-year plan to develop a bamboo nursery in collaboration with the Forest Right Act (FRA) 2006. Training and workshops can be organised to enhance bamboo and craft skills.

The second phase proposes innovation by merging traditional and modern technical knowledge using formal methods. During this phase, the SHG will work with a university-led short-term programme to support young women interested in creating bamboo crafts while incorporating technical knowledge.

The third phase entails developing marketing strategies based on demand. This involves collaborating with NGOs, government departments, villages, middlemen, and skilled artisans to create effective marketing and customer chains. The final phase focuses on sustaining the bamboo craft business by utilising new technologies and innovative strategies to expand the market for these products.

Proposed social interventions aim to develop innovative strategies for Gond women, ensuring their participation in bamboo craft activities and establishing sustainable livelihoods that empower them to engage in economic and political systems.

12.7 Conclusion

Bamboo craft is an important traditional occupation in which Gond women participate equally with men for survival. However, the work of Gond women is often overlooked, affecting their decision-making power at the family level regarding bamboo cutting, crafting, and selling processes. There is a need to promote community values and recognise Gond women's traditional knowledge and skills to address this issue. The Gond community values equality, and efforts should be made to create new opportunities for women that align with these values. A community-based intervention that enhances Gond women's potential, skills, and knowledge can lead to greater agency for women within and outside the community, promoting gender equality and equal opportunities.

References

Bain WK (2022) Traditional wisdom and practices involved in bamboo based crafts of the Lepcha community of North Sikkim—a case study from Dzongu Reserve Area. J Multidiscip Stud Archaeol 9:608–630

Baral SK (2014) Socio-economic empowerment of women by developing market potential of bamboo products in Nepal. INBAR, Beijing

Bayraktaroğlu S (2020) Actors of rural innovation: bamboo craft initiatives from Northeast India. In: Anderson LS (ed) Rural areas: an overview. Nova Science Publishers, New York, pp 145–190

Biberman J, Bajpai N (2020) India and the SDGs: towards a new Indian model of information and communications technology-led growth and development. Center for Sustainable Development, Earth Institute, New York

Carastathis A (2014) The concept of intersectionality in feminist theory. Philos Compass 9(5):304–314. https://doi.org/10.1111/phc3.12129

Chambers R, Conway G (1991) Sustainable rural livelihoods: practical concepts for the 21st century. IDS Discussion Paper 296. Institute of Development Studies, Brighton

Crenshaw K (1992) Mapping the margins: intersectionality, identity politics, and violence against women of color. Stanford Law Rev 43(6):1241–1299

Dasgupta S (2018) Adivasi studies: from a historian's perspective. Hist Compass 16(10):e12486. https://doi.org/10.1111/hic3.12486

Kabeer N (1999) Resources, agency, achievements: reflections on the measurement of women's empowerment. Dev Change 30(3):435–464. https://doi.org/10.1111/1467-7660.00125

Kabeer N, Narain N, Arora V et al (2023) Group rights, gender justice, and women's self-help groups: exit, voice, and loyalty in an indigenous community in India. Soc Philos Policy 40(1):103–128. https://doi.org/10.1017/S0265052523000390

Khan S, Hasan Z (2020) Tribal women in India: the gender inequalities and its repercussions. Int J Res 9(9):8–17

Krishna S (2001) Gender, tribe and community control of natural resources in North-East India. Indian J Gend Stud 8(2):307–321. https://doi.org/10.1177/097152150100800210

Krishna S (2003) Looking beyond academia. Econ Polit Wkly 38(31):3218–3219

Oestreich JE (2018) SDG 10: reduce inequality in and among countries. Soc Altern 37(1):34–42

Sengupta M, Harris LM (2022) Interrogating differences: intersectionality and participatory livelihood development in the upland forest of Tripura (Northeast India). Geoforum 130:59–68. https://doi.org/10.1016/j.geoforum.2022.02.002

Shaik RB (2016) Adivasi women in Southern India and their cultural oppression in contemporary India—a case study. Proc Indian Hist Congr 77:1065–1072

Xaxa V (2016) Tribes and Indian national identity: location of exclusion and marginality. Brown J World Aff 23(1):223–237

Street Gangs as a Pathway for Young Boys in the Western Cape, South Africa

13

Heidi Sauls and Robyn Singh-Adriaanse

Abstract

South Africa was under apartheid rule from 1948 to 1994, which allowed for the unequal development of different racial groups in the country. In response to this inequity, those who have been disadvantaged have found alternative ways to cope with the deprivation they experience. One such pathway is young boys joining street gangs. This chapter focuses on the findings of an ethnographic study that aimed to gain deeper insight into the experiences of Black African and Coloured boys (10–16 years old), who awaited trial in a Child and Youth Care Centre (CYCC) located in the Western Cape, South Africa. The research methods employed individual interviews, focus group discussions, and participant observation, with the young boys as the key participants. Based on the collected data, the boys' experiences and narratives revealed how intergenerational poverty shaped their lives, leading them to seek social, economic, and physical protection by joining street gangs, which often resulted in criminal and violent experiences. To mitigate this pathway for young boys, this chapter proposes that Haines and Case's *Child First* philosophy and approach be introduced and adopted as a social intervention to support and strengthen the ethos and objectives of the Child Justice Act (No. 75 of 2008) in South Africa.

Keywords

Street gangs · Juvenile justice · Child and Youth Care Centre · Intergenerational poverty · Youth violence

13.1 Introduction

Reflecting on South Africa's history, Sustainable Development Goal (SDG) 10—Reduced Inequalities—holds particular relevance. From 1948 to 1994, the apartheid regime institutionalised the separate development of different racial groups, embedding structural inequalities that continue to shape the present. Despite the passing years, remnants of this unequal system persist, as revealed in a World Bank report (Sulla et al. 2022), which identified South Africa as the most unequal country in the world, with race remaining a key determinant of inequality—where 80% of the nation's wealth is concentrated in the hands of just 10% of the population. Within this context, individuals with limited access to various forms of capital are more vulnerable (Bourdieu and

H. Sauls (✉) · R. Singh-Adriaanse
Department of Police Oversight and Community Safety, Western Cape Government, Cape Town, South Africa
e-mail: Heidi.Sauls@westerncape.gov.za

R. Singh-Adriaanse
e-mail: Robyn.Singh-Adriaanse@westerncape.gov.za

N. V. Roman et al. (eds.), *Social Inequalities Within and Among Countries*, Sustainable Development Goals Series, https://doi.org/10.1007/978-3-032-08741-6_13

Wacquant 1992).[1] Their marginal social positions often compel them to explore alternative, and at times socially marginalised, strategies in an effort to transform their daily experiences of deprivation.

This chapter presents the findings of a qualitative research study conducted to gain deeper insight into the lived experiences of Black African and Coloured boys (aged 10–16) who were awaiting trial in a Child and Youth Care Centre (CYCC)[2] in the Western Cape, South Africa. Drawing on the boys' narratives, the study highlights how their social positions and deprivation shaped their lives, often leading to involvement in crime, violence, and street gangs. The findings also point to significant gaps in the implementation of the Child Justice Act (CJA), No. 75 of 2008. In response, this chapter proposes the adoption of the *Child First* philosophy and approach as a social intervention to support and strengthen the implementation of the CJA.

13.2 Literature Review

Post-apartheid South Africa has yielded a mixture of opportunities and disappointments (Meintjies and Hall 2008). Although the democratic transition was premised on the promise of equal opportunity, stark inequalities persist. As noted in the 2022 World Bank Report, "Inequality in Southern Africa: An Assessment of the Southern African Customs Union", South Africa remains the most unequal country in the world. This disparity is particularly evident amongst children: whilst a minority grow up in conditions of extraordinary wealth, the majority face poverty and lack access to basic needs and essential services (Barnes et al. 2017). In response to such deprivation, some children exercise agency by engaging in criminal activities—becoming both perpetrators and victims shaped by their social circumstances. Understandings of their behaviour must therefore be grounded within this context that includes their daily exposure to violence—on the streets, in schools, and within the home (Lockhat and van Niekerk 2000; cf. Reynolds 1989). However, the ways in which children's behaviour reflects the country's past and present social issues remain unclear (cf. Abbink 2005: 18).

Moreover, extensive effort has gone into promoting restorative justice practices in the country due to the growing recognition that creating a supportive and rehabilitative environment is critical for children in conflict with the law. However, despite these proactive measures, there is still a need for continued advocacy and collaboration between stakeholders to ensure that child offenders receive the necessary support and opportunities, as well as for the root causes of juvenile delinquency to be addressed.

International scholars such as Rodgers and Jensen (2014) have sought to gain deeper insights into the dynamics of street gangs in various contexts. Similarly, in South Africa, qualitative studies have aimed to understand the complex factors underlying gang involvement. Hesselink and Bougard (2020) also employed the push–pull framework to explore the compounding risk factors and pathways that lead young males towards gang affiliation and violence, drawing on the narratives of incarcerated youth gang members. Childhood experiences reported by these young gang members, as push factors for youth gang involvement, included poor school commitment and low educational levels; poverty, abuse, and neglect; having a family member involved in criminal behaviour and gang involvement, or substance abuse; poor parental bonds, involvement, and supervision; family conflict and troubles within the home environment; truancy; poor self-restraint; and low self-confidence. Factors that drew youth into gang involvement included, for example, gangs serving as a substitute for absent families, gangs meeting survival needs, and high-crime, gang-infested neighbourhoods (Hesselink and Bougard 2020). Gang

[1] Bourdieu's Capital Theory states that individuals own various forms of capital that can shape their social positions but also influence social behaviour. Bourdieu refers to three forms of capital: economic, social, and cultural.

[2] A Child and Youth Care Centre (CYCC) is a form of residential care that accommodates more than six children and is registered with the Department of Social Development. Several interventions are rendered to improve the well-being of the children and assist their reintegration into society (Republic of South Africa [RSA] 2005). See Chapter 13, Section 191 of the Children's Act 38 of 2005.

membership can therefore be understood as an outcome of youth pursuing their unmet needs, driven by the conditions faced by young people in low-income settings (Pinnock and Pinnock 2019), a phenomenon also observed and supported in international studies (Hennigan and Spanovic 2020; Bellair and Browning 2020).

In the South African context, the state has traditionally approached gangs and gang-related violence with a reactive strategy, emphasising policing and the criminal justice system (NICOC Task Team 2015). However, more recently, there has been a call and shift for a broader, more holistic, approach to address the challenge of gangs. For instance, Viltoft (2022) argued for sustained anti-gang strategies and interventions that address structural obstacles and inequality given the unique context of South Africa and the Western Cape, in particular.

13.3 Research Methodology

The findings presented in this chapter emerged from an ethnographic study that employed various child-centred qualitative research methods. The research methods included participant observation, semi-structured interviews, and focus group discussions. The study explored the young boys' life histories, experiences, and perceptions in and out of the CYCC. A limitation of this study is that the findings cannot be generalised to young boys in the province. Also, it is possible that desirable answers were provided by the participants and could influence the findings.

13.3.1 Research Sites and Participants

The research site was a CYCC[3] for boys in conflict with the law. The participants, aged between 10 and 16 years, originated from urban, semi-rural, and rural areas within the Western Cape. Participant recruitment was facilitated through a combination of convenience and snowball sampling methods. Convenience sampling enabled the inclusion of accessible and willing participants (Teddlie and Yu 2007), whilst snowball sampling allowed participants to refer and connect the researcher to other potential participants within the centre.

13.3.2 Data Collection and Analysis

The qualitative research instruments included semi-structured interview schedules. The questions were presented in English and Afrikaans. Research questions were designed for individual interviews with the boys as key informants. Questions were also compiled for the focus groups. All engagements with the boys were recorded and transcribed. Multiple levels of data analysis were then conducted. The first level of analysis broadly identified themes that emerged from the data, whilst the second level further refined these themes. The main themes recognised from the data were then analysed.

13.3.3 Ethical Considerations

Ethical approval was granted by the Western Cape Department of Social Development. The boys were informed about the purpose of the study and invited to participate on a voluntary basis. They were assured that they could withdraw from the study at any time without repercussions. Furthermore, they were informed that their involvement would remain confidential and anonymous. A designated space was provided for the interviews, and during the analysis, the names of the participants were replaced with pseudonyms to protect the participants' identities. Finally, caregivers were required to provide consent in addition to the young boys' assent forms.

[3] Due to confidentiality, the name of the CYCC will not be mentioned.

13.3.4 Limitations of the Study

The findings of the study are based on a specific group of boys in one CYCC. This limits the extent to which the results can be generalised to all boys in conflict with the law across the country. Secondly, convenience and snowball sampling possibly excluded boys who were less willing to participate, hereby potentially skewing the data. Finally, the study relied heavily on the participants' narratives which might have been influenced by memory biases and social desirability.

13.4 Findings and Discussion

13.4.1 Impact of Intergenerational Deprivation on the Boys' Lives

At the time of the study, the boys accommodated in the CYCC were profiled as young, Black African, and Coloured individuals raised in fragile families. Their families were historically disadvantaged, and, as a result, the boys experienced multiple forms of deprivation (Barnes et al. 2017). These challenges extended beyond financial issues and included low educational backgrounds and substance abuse, amongst others. The family's social position and overall deprivation inadvertently shaped the boys' daily lives and experiences. For instance, the boys frequently described how their families' financial struggles not only hindered their ability to meet their basic needs but also influenced their behaviour.

A 14-year-old participant said: "I steal for myself so that I can eat", whilst another participant shared, "I can guarantee that they [the other participants] wouldn't know what they would eat at night, that includes me. It's going to be whatever my [single] mother can get… Everything is hard". Several boys also described how their caregivers' financial instability and poverty placed strain on family members and resulted in conflict and violence amongst them.

13.4.2 Masculinity, Violence, and Vulnerability

The boys' narratives revealed the violent and unsafe environments they navigate, resulting in their experiences of poly-victimisation across multiple facets of their lives since early childhood. The forms of violence they encountered included physical assault, bullying, robbery whilst walking to and from school, and sexual violence. The boys also recounted how gang members victimised them, even though they were not gang members themselves. "*… And that's how it [deciding to join a gang] starts*", explained one of the boys. The interviews clearly highlighted the boys' ongoing victimisation in the spaces they inhabit. It was evident that the boys, who are in conflict with the law, are equally at risk of being victimised, yet their victimisation remains unacknowledged and unaddressed.

13.4.3 Street Gangs as a Pathway for Young Boys

Based on the interviews, it was revealed that the boys employed alternative pathways to fulfil their basic needs, which included less socially acceptable efforts such as gang membership. The boys described how being a gang member allowed for some of their needs to be met, despite the risk.

> [Our] Families don't have food. But you know that you will get [a meal] from the gang… you get the tattoo and then you belong to a family. Now, you have a family, so you can go eat there.

The boys claimed that the gangs are aware that families are vulnerable and then attract children with material goods.

> They pull you in with stuff that you want, that you know you won't get at your house… so you will do anything to get it. Now you go onto the wrong path [join the gang and commit crimes] and do what they tell you to do.

Furthermore, the boys explained that if they do not act, their needs will not be met.

> ...every day we must stand outside [on the streets], because if you lay at home, you are going to get nothing [no money, food, etc.] ... then we're not going to survive.

Street gangs therefore become a viable option for the boys because the gang members work together to accumulate resources. Gang affiliation also functions as a response to the unsafe environments the boys must navigate. However, this protection comes at a significant cost. Participants described feeling coerced into repaying favours, leading to entrapment.

> ... you ask them if they can help you out and then, you owe them. Now you're in it [the gang]. You can't get out.

Whilst gang membership may offer a degree of protection from daily physical threats, it simultaneously increases the risk of exposure to gang violence and even death.

13.5 The Intersection of Gender, Race, Deprivation, and Street Gangs

Crenshaw's (1989) intersectionality framework was used as an analytical tool to interpret the collected data. Intersectionality examines how overlapping social identities shape experiences of privilege, discrimination, and oppression. It emphasises that interconnected social categories lead to unique forms of discrimination based on multiple identities, such as gender, sexuality, and class. This perspective argues that various oppressions, such as those based on race and gender or sexuality and nationality, coexist to produce injustice (Collins 2000). Additionally, intersectionality explores intra-group nuances (Jordan-Zachery 2006) and how privilege and oppression can shift in different contexts (Kelly et al. 2021).

This framework served as an analytical tool to organise and interpret how the intersecting social identities of young boys, in conflict with the law, shape their lives and produce social and structural vulnerabilities that increase the risk of gang involvement and criminal behaviour. It aims to amplify the perspectives of young, marginalised boys and highlight the complexity of their experiences and behaviours, which are often too simply viewed through a criminal lens.

Based on the data, it was clear how the overlapping of intergenerational and persistent poverty, race, and gender produced different forms of injustice for the boys (Collins 2000). The intersection of their racial profiles, alongside the historical poverty of their families and, in turn, the boys themselves, translated into limited opportunities and resources. Gender compounded their experiences of inequality and discrimination, as their evident vulnerability to violence, lack of opportunities, and need for various forms of support often went unacknowledged. In response to their needs and vulnerability being overlooked, the boys exercised agency and created alternative ways to respond to their marginalisation, deprivation, and lack of support.

The intersection of poverty, masculinity, race, and social exclusion led the boys to join street gangs and engage in criminal activities, ultimately resulting in their arrests. Gang membership and participation can therefore be understood as the boys exercising their agency and attempting to address the poverty and lack of safety they experience. The street gangs and related activities therefore became a vehicle for the boys to accumulate different forms of capital, such as financial wealth and social power, in a context where this is limited (Boden 2019).

13.6 The Social Intervention: 'Child First'

The CJA places a legal obligation on the justice system to consider a child's individual circumstances and background when determining outcomes for children in conflict with the law. Although the CJA mandates a child-centred approach, multiple studies have shown that it has not been implemented as intended—with unclear reporting, resource constraints and systemic barriers hindering effective application of its core provisions. In addition, not all stakeholders dealing with children are aware of the CJA or adopt its restorative ethos (Chikukula 2022). In response to these challenges, Case and Haines' (2014) '*ChildFirst*' approach is proposed as a means to support and strengthen both the underlying philosophy and the practical application of the CJA.

Case and Haines introduced the 'Child First, Offender Second' approach in 2014, which was more recently referred to as *Child First* because 'offender second' continued to label children, who are in conflict with the law. This model prioritises positive behaviour and outcomes for children in conflict with the law. It challenges the predominant negative, retrospective, and offender-focussed methods in youth justice and aims to avoid labelling children as offender;, instead addressing the systemic factors influencing their behaviour (Case and Haines 2014). This approach also encourages collaboration with the children themselves and emphasises prevention, whilst striving to reduce the stigma attached to children who have had contact with the criminal justice system. It particularly encourages those working in youth justice services to 'prioritise the children's best interest' (Case and Haines 2021). Therefore, Child First has a dual purpose: to identify and address the influences on offending, whilst also identifying and promoting factors that support children's desistance from offending, thereby helping them to move into pro-social, positive behaviour (Case and Browning 2021).

The strategic goal of the *Child First* approach is grounded in evidence-based principles that inform a comprehensive model of practice (Case and Browning 2021). According to Case and Browning (2021), the approach is built on four core tenets: (1) viewing children as children; (2) fostering a pro-social identity to promote positive child outcomes; (3) collaborating with children; and (4) promoting diversion, thereby ensuring a childhood free from justice system involvement.

Raising awareness amongst all professionals working with children about the *Child First* approach—and how the intersection of various social categories can heighten a boy's vulnerability to gang involvement—may enable earlier identification of the challenges faced by boys and their families. This, in turn, allows for timely referrals to relevant stakeholders and interventions, serving as a preventative measure to reduce their contact with the criminal justice system. Moreover, this approach supports the diversion of more boys away from the justice system, helping them avoid its potentially harmful consequences at a young age. Ultimately, it reduces the risk of future social exclusion and marginalisation, which can hinder employment opportunities and perpetuate cycles of disadvantage.

The prioritisation of the child's best interests is, therefore, a constructive and future-focussed approach that promotes the development of a pro-social identity for sustainable desistance (Case and Browning 2021). Supporting this guiding principle of *Child First* within the existing policy and implementation of the CJA will contribute to strengthening a more socially inclusive awareness and response when dealing with children before they engage in criminal and violent behaviour.

Finally, Case and Browning (2021) primarily focus on youth justice practitioners who deal with children in conflict with the law. They suggest that efforts to strengthen these principles should be continuous. The authors of this chapter, however, propose that an awareness and implementation of the Child First approach and its principles be extended beyond the youth sector and criminal justice. Additional key stakeholders, at an operational, policy and strategic level should also be made aware of how categories such as poverty, gender, and race can influence behaviour, including criminal and violent behaviour. Therefore, the authors suggest that educators in the school system and early childhood development (ECD) centres, for example, be educated on the Child First approach and how the intersection of different social categories can shape young boys' actions. Practitioners and policy developers should be trained to identify their own prejudices regarding Black African and Coloured boys from less privileged families and communities. Challenging existing biases against these young boys can lead to more empathetic interventions and day-to-day interactions, whilst also facilitating access to appropriate support for the child and their family. In doing so, it becomes possible to disrupt the trajectory that may lead a boy towards gang involvement and arrest.

13.7 Conclusion

Adoption of the *Child First* approach amongst practitioners and policy developers beyond youth justice will reinforce the CJA, empower young boys in conflict with the law and promote inclusivity. These efforts will acknowledge the boys' vulnerabilities that can lead them to join gangs, whilst also recognising that gang membership and its activities represent a form of agency shaped by socio-economic and cultural contexts. This understanding allows for preventative and supportive interventions apart from the criminal justice system. By grasping these influences, the objectives of SDG 10.2 to promote social, economic, and political inclusion will be strengthened.

References

Abbink J (2005) Being young in Africa: the politics of despair and renewal. In: Abbink J, Kessel I (eds) Vanguard or vandals: youth, politics and conflict in Africa. Brill, Leiden, pp 1–34

Barnes H, Wright G, Noble M, Dawes A (2017) Poverty and inequality in South Africa: evidence from the national income dynamics study. J South Afr Stud 43(1):1–17

Bellair PE, Browning CR (2020) Contemporary disorganization research: an assessment and further test of the systemic model of neighborhood crime. J Res Crime Delinq 57(6):789–823

Boden T (2019) Young people, gangs and street-based violence. Aggress Violent Beh 48:101337. https://doi.org/10.1016/j.avb.2019.07.002

Bourdieu P, Wacquant LPD (1992) An invitation to reflexive sociology. University of Chicago Press, Chicago

Case S, Haines K (2014) Children first, offenders second—positive promotion: reframing the prevention debate. Youth Justice 15(3):226–239

Case S, Haines K (2021) Abolishing Youth justice systems: children first, offenders nowhere. Youth Justice 21(1):3–17

Case S, Browning A (2021) Child first justice: the research evidence-base [full Report]. Loughborough University. https://hdl.handle.net/2134/14152040.v1

Chikukula H (2022) An implementation analysis of the Child Justice Act 75 of 2008 with specific focus on child justice courts in South Africa (Masters thesis). University of KwaZulu Natal.

Collins PH (2000) Black feminist thought: knowledge, consciousness, and the politics of empowerment, 2nd edn. Routledge, New York

Crenshaw K (1989) Demarginalizing the intersection of race and sex: a black feminist critique of antidiscrimination doctrine, feminist theory and antiracist politics. Univ Chic Leg Forum 1989(1):139–167

Hennigan K, Spanovic M (2020) Gang dynamics through the lens of social identity theory. In: Esbensen F-A, Maxson CL (eds) Youth gangs in international perspective: results from the Eurogang Program of Research. Springer, pp 127–148

Hesselink A, Bougard NB (2020) Risk factors associated with youth gang involvement: an exploratory criminological case study analysis. J Psychol Afr 30(5):459–465. https://doi.org/10.1080/14330237.2020.1821314

Jordan-Zachery JS (2006) Am I a black woman or a woman who is black? A few thoughts on the meaning of intersectionality. Polit Gend 2:254–263

Kelly C, Kasperavicius D, Duncan D, Etherington C, Giangregorio L, Presseau J, Sibley KM, Straus S (2021) 'Doing' or 'using' intersectionality? Opportunities and challenges in incorporating intersectionality into knowledge translation theory and practice. Int J Equity Health 20, Article 187. https://doi.org/10.1186/s12939-021-01509-z

Lockhat R, van Niekerk A (2000) South African children: a history of adversity, violence and trauma. Ethn Health 5(3–4):291–302. https://doi.org/10.1080/713667462

Meintjes H, Hall K (2008) Demography of South Africa's children. In: Pendlebury S, Lake L, Smith C (eds) South African Child Gauge 2008/2009. Children's Institute, University of Cape Town, Cape Town

NICOC Gangsterism Task Team (2015) National Anti-Gangsterism Strategy. NICOC, Pretoria

Pinnock D, Pinnock R (2019) Strategic roadmap towards implementation of the National Anti-Gangsterism Strategy in the Western Cape: provincial response to the National Anti-Gangsterism Strategy (NAGS) 2019, South African Drug Policy Initiative. http://www.sadpi.org/wp-content/uploads/2019/10/Roadmap-for-Safety-Report-.pdf

Republic of South Africa (RSA) (2005) Child Act 2005, No. 38. Government Printer, Pretoria

Reynolds P (1989) Childhood in crossroads: cognition and society in South Africa. David Phillip, Cape Town

Rodgers D, Jensen S (2014) The gang phenomenon in global perspective. In: Hazen JM, Rodgers D (eds) Global gangs: street violence across the world. University of Minnesota Press, Minneapolis, pp 1–25

Sulla V, Zikhali P, Cuevas PF (2022) Inequality in Southern Africa: an assessment of the Southern African customs union. World Bank, Washington, DC

Teddlie C, Yu F (2007) Mixed methods sampling: a typology with examples. J Mixed Methods Res 1(1):77–100

Viltoft CD (2022) Deconstructing gangsterism in South Africa: uncovering the need for gender-sensitive policies. Cosmop Civ Soc: Interdiscip J 14(3):1–14. https://doi.org/10.5130/ccs.v14.i3.7985

Left Out but Not Left Behind: Community-Based Kindergarten Social Intervention for Children of Plantation Workers in Kedah, Malaysia

14

Veronica Anne Retnam

Abstract

Children in several pockets of rural and urban poor areas in Peninsular Malaysia have historically been excluded from kindergarten programmes. Whilst a limited number in some impoverished communities had access, these programmes were not context-sensitive to the realities of plantation life. Such gaps led to social intervention through community-based kindergarten efforts to bridge the inequalities and to innovate programmes to address the intersectionality of these inequalities Community-based kindergarten teachers involved in these efforts were identified and trained by several community organisations. Pachaiammal, who worked in a rubber plantation, was trained by the Persatuan Guru-Guru Tadika Malaysia (PGT) beginning in 1989. PGT empowered her to provide holistic care and development for children in her community. This chapter identifies and analyses Pachaiammal's journey to effect change through social interventions pioneered by PGT. Using a qualitative approach with a case study design, and employing open-ended and semi-structured questions, a sample of five participants was asked to critically reflect on their kindergarten experience. The participants included Pachaiammal, her two former kindergarten students (now adults), and two parents. The interviews were conducted in Tamil, translated into English, and then transcribed. The findings reveal how Pachaiammal transitioned from solely providing kindergarten access for ten years without parental support to being empowered to secure the active involvement and participation of parents and the community in the holistic care and development of their children. This change was realised through a collaboration that PGT developed with concerned groups. Consequently, those children, as adults with community values, became involved in the Malaysian economy and benefited from its development. The social intervention of Pachaiammal, who effectively applied her PGT training, was aligned with Sustainable Development Goal (SDG) 10.2, which collaboratively identified and addressed the intersectionality of the inequalities faced by marginalised communities.

Keywords

Community-based kindergartens · Holistic · Collaboration · Marginalised

14.1 Introduction

Rubber plantations in Peninsular Malaysia are a legacy of British colonial policy, which brought indentured labour from Britain's other colony, India, at the end of the nineteenth century (Jain

V. A. Retnam (✉)
Seremban, Negeri Sembilan, Malaysia
e-mail: varetnam16@gmail.com

N. V. Roman et al. (eds.), *Social Inequalities Within and Among Countries*, Sustainable Development Goals Series, https://doi.org/10.1007/978-3-032-08741-6_14

1970). In terms of education, the Tamil plantation schools introduced by the British functioned as agents of social reproduction, as many children dropped out of primary school to assist their parents and eventually entered the labour force (Marimuthu 1993). In response to the community's needs across various regions of Peninsular Malaysia, several individuals and organisations took action. Persatuan Guru-guru Tadika Malaysia (PGT) was one such organisation.

As a community-based entity, PGT emerged through the efforts of the Consumer Association of Taiping, a small town in the State of Perak surrounded by an extensive hinterland of rubber plantations. PGT's initiatives within plantation communities gained the attention of the Tamil media, particularly as the association's appeals to integrate estate kindergartens into state-run rural kindergarten systems were unsuccessful (Manickam 2009). Ironically, as Arokiam and Raja (2019) note, the realities of estate life that persisted into the 1990s were exacerbated in the twenty-first century. A striking example of the enduring impact of these structural realities is the over-representation of Malaysian Indian male youth and young adults in gangsterism and crime (Ravichandran and Tharshini 2023). This trajectory is connected to the dominance of the Indian labour class in rubber plantations. Approximately 85% of Indian labour served British plantation capital—a pattern that persisted after Malaya's independence. As socio-economic development reshaped land use, many Indian labourers were displaced, with migrant workers increasingly filling their roles (Arokiam and Raja 2019). Related to Sustainable Development Goal (SDG) 10, the intersectionality of ethnicity, class, economic sector, age, gender, culture, location, and politics arises from the historical link of plantation labour serving capital.

A new era emerged for parents and the broader community through the initiative of a group of women from the plantation workers' community, supported by concerned individuals from outside the plantations. This collaboration marked the beginning of PGT. This group responded to pressing concerns about social inequalities in both access to and the quality of kindergarten education for children of plantation workers and those living in urban poor areas across several regions of Peninsular Malaysia. PGT's efforts resulted in targeted social interventions that addressed the intersectionality of these challenges, aligning with SDG 10.2. The interventions included training kindergarten teachers and empowering parents and communities to take active responsibility for the holistic development of children aged four to six. This chapter presents a case study of Pachaiammal Muniandy, highlighting the social interventions undertaken by PGT.

14.2 Contextualising the Case Study

This case study was conducted in Bukit Sidim Estate, located in the state of Kedah in northern Peninsular Malaysia. The nearest town, Kulim, is 25 km away. The estate is owned by the National Land Finance Cooperative Society (NLFCS), an initiative founded by Tun V. T. Sambanthan, one of Malaysia's founding fathers. The workers who relocated to Bukit Sidim Estate came from nearby estates that had previously been under British ownership. Their narratives closely reflect the themes identified in earlier scholarship on plantation history (Jain 1970) and more recent analyses (Arokiam and Raja 2019). Although the state provided kindergarten education in rural areas, the children of plantation workers were excluded, as plantations were considered private property (Manickam 2009). This exclusion persisted despite the significant contribution of rubber and palm oil to Malaysia's foreign exchange earnings up until the 1980s.

Due to poor quality and other contributing factors, plantation schools recorded the highest dropout rates in Malaysia (Malaysia. Kementerian Pelajaran 1973).[1] These outcomes were largely the result of unintended consequences of state policies, which perpetuated intergenerational poverty and entrenched social inequalities. Nevertheless, in the absence of formal state support for

[1] The only comprehensive study on school dropouts conducted to date.

kindergartens in plantation areas, such institutions emerged through the efforts of individuals affiliated with consumer movements, development funding from foundations such as the Bernard van Leer Foundation (BVLF), and the Catholic Church's commitment to social justice through its teachings. The kindergarten at Bukit Sidim Estate stands as one of PGT's earliest community-based kindergarten initiatives.

14.3 Literature Review

The intersectionality of ethnicity, class, economic sector, age, culture, location, and other socio-economic and political realities surrounding PGT created fertile ground for community-based social interventions. Tesoriero (2010) describes community-based services as part of community development, emphasising the importance of deconstructing social structures to foster empowerment. A notable example is the work of the Centre for Health, Education, Training and Nutrition (CHETNA) in India, which empowers marginalised communities across several states (CHETNA India n.d.). In the area of early childhood care and education, CHETNA supports Anganwadi Centres, which are publicly funded early childhood development facilities.[2] Established in 1980 as a resource agency, CHETNA has since played a key role in grassroots development (CHETNA India n.d.). Similarly, the case study of Pachaiammal's community kindergarten effort uncovers intersectionality arising from the lived realities of marginalised children, teachers, parents, and community members, leading to social intervention as a response to social inequalities. This case aligns with the goals of SDG 10.2, which seeks to promote the social, economic, and political inclusion of all.

14.4 Methodology

Open-ended and semi-structured questions propelled this qualitative study, which used a case study design. The case study focuses on one individual, Pachaiammal Muniandy, who was selected due to her thirty-year commitment as a PGT community-based kindergarten teacher and PGT member. Interviews with Pachaiammal were conducted at her home after her kindergarten sessions. The interviews began with open-ended questions, whilst some were structured. Although the case study centres solely on Pachaiammal, other participants recalled and shared their memories of their involvement in Pachaiammal's kindergarten efforts. This research thus represents a retrospective study. The data collection involved various categories of participants, including Pachaiammal, former kindergarten attendees, and parents of the attendees, bringing together multiple perspectives on Pachaiammal's social intervention. All participants were of Tamil origin, with ages ranging from 21 to 74 years.

In addition to Pachaiammal, there were three female participants and one male participant. The interviews, conducted in Tamil, were digitally recorded, translated into English, and transcribed. The accuracy of the translation was verified by a Tamil-speaking grassroots activist with tertiary qualifications to ensure the original meaning was preserved. Findings were then categorised according to patterns related to programmes for teachers, children, and parents. These patterns were analysed according to themes and subthemes. The overlapping and interrelatedness between the emerging themes were also analysed in relation to social inequalities, intersectionality, and social interventions.

14.5 Findings

The findings focus on the teacher's background, social intervention programmes designed for their development, and initiatives for children and parents introduced by Pachaiammal. The key thematic areas that emerged were: programmes for teachers, programmes for children, and programmes for parents and the broader community.

[2] www.chetnaindia.org.

14.5.1 The Teacher's Background

When Pachaiammal began the kindergarten initiative in 1979 at Bukit Sidim Estate—her birthplace—she and other workers' families lived in rows of houses known as 'labour lines'. The estate community consisted primarily of workers of South Indian origin, alongside a number of Malay workers, totalling approximately 500–600 individuals at the time. Later, the National Land Finance Cooperative Society (NLFCS) built houses for workers in nearby Labu Besar,[3] prompting Pachaiammal's family, along with others, to relocate there.[4]

The Tamil worker community preserved many of the cultural practices that their forefathers had brought with them from their villages in South India, including caste-based male hierarchies and conservative religious practices. As the eldest of seven siblings, Pachaiammal began her formal education at the age of six, encouraged by her teacher, Mr. Subaya, who inspired her dream of becoming a teacher. After completing five years of public secondary education, she obtained the Malaysia Certificate of Education and chose to pursue a career in early childhood education.

In 1976, she began her training as a kindergarten teacher through the Educational, Welfare and Research Foundation (EWRF) in Kuala Lumpur, Malaysia. In 1979, she underwent further training at the Sri Avinashilingam Home Science College in Coimbatore, India, as part of a three-month programme. She was one of 15 girls selected and sponsored by Tun V. T. Sambanthan to be trained as kindergarten teachers for NLFCS-owned estates. Upon completion of her training, she returned to Bukit Sidim Estate to begin her teaching career.

Between 1979 and 1989, Pachaiammal used songs and stories guided by themes she learned during her short training in India. Although children had access to some form of early learning, she often felt inadequate and yearned for further knowledge and development. A significant turning point came in 1989 when she was selected for additional training through PGT. Reflecting on this opportunity, Pachaiammal passionately recalled:

> Their [PGT] intention was to ensure that no child from the estate should be left behind. Every child must get knowledge. They had that vision. PGT is Caring, Sharing, Loving. The teacher must not be someone teaching without any knowledge. (Pachaiammal)

14.5.2 PGT's Social Interventions for Teachers

Using a mind-mapping technique, Pachaiammal outlined how she benefited from PGT's six-level training programme, which was delivered over four years in three languages—Tamil, Bahasa Malaysia (the national language), and English. She emphasised that PGT's interactive and small-group training methodology, which followed brief theoretical sessions with hands-on learning, set it apart from her earlier experience with the EWRF programme. Reflecting on her training, Pachaiammal remarked that she met many girls like herself from other estates and formed lasting friendships. As she put it, '[The PGT training] helped me to upgrade myself' (Pachaiammal).

Pachaiammal explained how a 'hands-on' training programme taught her and her friends to use recyclable materials to create teaching aids. They began by collecting waste materials from shops in Taiping and later expanded their efforts to their neighbourhoods. Pachaiammal noted that she gained confidence as she progressed through the levels, culminating at Level Six in 1994, where she prepared teaching materials for six learning corners and developed accompanying workbooks.

Pachaiammal cherished the friendship and networking opportunities provided by PGT's meetings, where teachers could learn from and support each other in their districts. Some parents were also included. She noted that these meetings were rotated amongst estates four times a year or more to organise and coordinate community events such as the annual sports day, concert, women's day, and cultural events like *Pongal*—

[3]Though 1000 houses were planned, only 48 houses were built.

[4]The cooperative-built houses in the estate have more living space and basic facilities.

the Tamil harvest festival. Other religious and cultural festivals included Hari Raya Aidil Fitri, Chinese New Year, and Christmas. With funding from the BVLF from 1994 to 2006,[5] Pachaiammal gained knowledge and skills through PGT's training programme which was influenced by CHETNA. She subsequently became a PGT trainer in addition to her role as a kindergarten teacher. Remarking on her social interventions, she expressed, 'That is what I bring to the children and community'.

14.5.3 Programmes for Children and Parents

Programmes for children, both within and beyond the kindergarten setting, were supported through the active participation of parents and the wider community. Tarani, a 25-year-old medical student who attended Pachaiammal's kindergarten in 2005, reflected on her early learning experience:

> The first time I go, I remember I cry, then I started to ahh, started to…bond with the children there, then I started to study. Since it is a Tamil *Tadika* [kindergarten], I started to study Tamil first, then only I started to study in English and Bahasa Malaysia. (Tarani)

Tarani provided an example of how she and her friends learned the name of an orchid. Pachaiammal had turned it into a play, with children excitedly taking turns to place orchids in each other's hair whilst learning its pronunciation. Tarani reflected on her kindergarten experience:

> I realise it is once in a lifetime. Now it is hectic days. In kindergarten, we did not know [the word] 'tired'. We did playful things. Teacher was friendly with me and my friends. (Tarani)

Pachaiammal provided an example of how she conducted a class to prepare a budget for children before Deepavali,[6] a religious festival for Hindus. As the children had expectations of presents beyond their parents' means, she guided them to reflect on the salaries their parents earn and how those small salaries[7] needed to be 'stretched' for home expenses. 'I conducted a class on how to prepare a budget before Deepavali with the children. I had learned this from Bro. John'[8] (Pachaiammal 1/5/2021). As a result of that budget lesson, the children accepted what their parents could afford: usually a pair of affordable shoes.

Another former student, Kayal Villy Venkitaraman, 45, who is currently a primary school teacher and attended Pachaiammal's kindergarten from the age of four, commented on Pachaiammal's training over the years:

> In our time, we couldn't write sentences in Year One. We could read suku kata [syllables], words. But we have difficulties constructing a sentence on our own. Children who came after that had more books, and teacher used to teach more, and she had more worksheets too. These students were more advanced than us. There are dentists, homeopathy doctors, and lawyers. (Kayal)

Kayal's sharing resonated with Pachaiammal's appreciation of the materials provided through PGT:

> We never bought books. The books they [PGT] gave is what I used for the children. Secondly, they gave stationery. (Pachaiammal)

Another aspect of the programmes implemented to address social inequalities was targeted at parents and other members of the community. Pachaiammal was the only teacher for 30–40 children, like most other PGT teachers. The PGT programmes empowered teachers through community leadership training, according to Pachaiammal. She learned to write letters to government officials and to follow up in person in order to gain access to facilities like the local council field. In addition, Pachaiammal approached dignitaries such as the Member of Parliament to grace events like the district-level sports held in Kulim town.

She expressed gratitude to Dr Paul Sinnapan, who, together with his spouse, Papathiammal, empowered teachers to break free from the limi-

---

[5]BVLF funded PGT to sustain its programmes through the Asian Institute for Early Childhood Care and Education (AIECCE).

[6]It is also known as Diwali.

[7]Rubber tappers' wages were below the poverty line for Peninsular Malaysia in the 1980s and 1990s.

[8]Brother John D'Cruz, 83 years old, is a La Salle religious brother who brought alternative education to PGT.

tations imposed on them by society. Through these self-development programmes, Pachaiammal stated that she and her friends learned to organise activities on a scale beyond the capabilities of a single teacher or the community. Parents marched with their children, unlike in other kindergartens.

The parents of her former kindergarten children were grateful to Pachaiammal. Tarani's mother, Gunasheela, 45 years old and a factory worker on shift work, depended on Pachaiammal to 'take care' of her children whilst she was away from home for long hours. Another parent, Maraya, 74, who dropped out of school at the age of nine, commented on Pachaiammal's abilities:

> She was the only one teaching the children with a lot of care, a lot of interest. That is how my children got the courage [tunichal] and moved on from the first class to the next [Year One to Year Two]. She was a good role model for the children, especially given today's situation. She is still a teacher. (Maraya)

As parents, they relied on Pachaiammal for the basic learning she provided through the kindergarten. Maraya reflects on Pachaiammal's efforts and commitments:

> Yes, she has worked hard, she has sacrificed, not got married. She has done all these for the children. If she had thought, she too could have had a family. We must appreciate her. (Maraya)

As she reflected on her work with the community, Pachaiammal felt a sense of fulfilment:

> I did this job truthfully and with love. Till now, I am 64, God has taken care of my wellbeing [*arokiam*]. Another thing: my children, when they see me, say 'I have come up because of her; if I am doing all these, my teacher is the reason.' I have achieved my goal a lot. Many people have said that I have done this job without expectations. Till now I have not taken any medicine for this *arokiam*. I have not been admitted in hospital. (Pachaiammal)

14.6 Discussion

The social intervention implemented by PGT addressed the intersectionality of ethnicity, class, economic sector, age, gender, culture, location, and politics, enabling children from the plantation community to overcome inequalities through Pachaiammal's community-based kindergarten initiatives. Participant testimonies indicate that these children successfully transitioned from the kindergarten to the Tamil primary school within the plantation and subsequently to secondary school in town. As a result, the Bukit Sidim Estate Tamil school no longer functioned as an agent of social reproduction—a concern previously raised by Marimuthu (1993) regarding plantation Tamil schools. Through training provided by PGT, Pachaiammal was able to promote holistic child development, equipping children with both social and academic skills essential for school readiness. Her efforts also extended equal early learning opportunities to children who had been excluded from the state kindergarten programme, as noted by Manickam (2009).

PGT's social intervention, led by Pachaiammal, received active support from parents and the wider community—not only in affirming the role of the teacher but also in fostering shared responsibility for the care and education of their children. Through the social intervention programmes, she learned and adapted via PGT, Pachaiammal brought dignity to the rubber-tapping community. Her efforts demonstrated that parental support and meaningful involvement are essential components in promoting high-quality parental engagement in early childhood education (Kambouri et al. 2022). Whilst Kambouri et al. (2022) proposed a theoretical model, Pachaiammal successfully implemented a community-based programme that attracted sustained parent and community involvement through a range of inclusive activities, such as sports events and concerts. As a result, the social intervention positively impacted not only the children from the community but also the teachers and parents.

14.7 Conclusion

The findings of this study indicate the impact of Pachaiammal's formation through PGT as a versatile community-based kindergarten teacher in creating a joyful environment that enables children to learn. Her empathy for children, stemming from her own social reality, translates into a commitment to their care and education. Programmes involving parents and the community have ensured their dedication to their children's holistic development. Pachaiammal's contribution to her community in addressing inequalities in educational opportunities represents a significant social intervention. I make this claim based on the fact that Pachaiammal's former kindergarten students, who were excluded from state kindergarten programmes, were not left behind in Malaysia's development; they became gainfully employed. Pachaiammal stated that finance did not matter; her passion and commitment to the children and the community did. Her social intervention through a nascent teacher formation programme has the potential to reduce inequalities in education.

References

Arokiam S, Raja SS (2019) Path dependent development of Indian plantation labourers in Malaysia: Unfolding the historical events in understanding their socioeconomic problems. Times Academic Press, Singapore, pp 79–102

CHETNA India (n.d.). https://www.chetnaindia.org. Accessed 13 July 2024

Jain RK (1970) South Indians on the plantation frontier in Malaya. University of Malaya Press, Kuala Lumpur

Kambouri M, Wilson T, Pieridou M et al (2022) Making partnerships work: proposing a model to support parent-practitioner partnerships in the early years. Early Child Educ J 50(4):631–669

Malaysia. Kementerian Pelajaran (1973) Kajian keciciran: drop out study. Dewan Bahasa dan Pustaka, Kuala Lumpur

Manickam JR (2009) The Malaysian Indian dilemma: The struggle and agony of the Indian community in Malaysia. Nationwide Human Development and Research Centre, Kuala Lumpur

Marimuthu T (1993) The plantation school as an agent of social reproduction. In: Sandhu KS, Mani A (eds) Indian communities in Southeast Asia. Times Academic Press, Singapore

Ravichandran S, Tharshini NK (2023) Gang membership among Malaysian Indian young adults. Int J Acad Res Bus Soc Sci 13(7):1325–1344

Tesoreiro F (2010) Community development: community-based alternatives in an age of development. Pearson Australia, Melbourne

Promoting Gender Equality: Working with the Girl-Child from the Margins to the Centre

15

Maria van Staden

Abstract

Adolescent girls in farming communities in the Western Cape, South Africa, face intersecting inequalities rooted in gender, poverty, geography, and cultural norms that limit their access to education and empowerment. This action research study explored how contextually tailored non-formal education programmes could serve as transformative interventions for this marginalised population. An intersectional lens informed the design and implementation of this non-formal educational programme for rural girls. The programme involved a residential workshop focussed on building life skills, gender rights awareness, and empowerment specifically for adolescent girls in the Western Cape's farming communities. Interviews and observations with participants, parents/guardians, and mentors assessed the impact of the intervention through an intersectional lens. Findings revealed that the programme enhanced participants' confidence, communication abilities, and self-expression, improving family relationships and challenging patriarchal gender norms that silenced their voices. The study highlights the significant contribution of mentorship in fostering growth and development for girls beyond what their families provide. The girls shared their acquired knowledge on sexual health, financial literacy, and more within their communities, positioning them as catalysts for broader intergenerational dialogue around gender and other inequalities. The study concludes that non-formal educational models can equip marginalised rural girls in the Western Cape with practical tools to navigate intersecting barriers whilst fostering agency and inclusion. This work contributes to an intersectional understanding of how grassroots non-formal interventions can advance progress towards the United Nations Sustainable Development Goals, particularly those related to gender equality and the reduction of inequalities.

Keywords

Girls · Non-formal education · Gender · Inequality · Intersectionality

15.1 Introduction

Gender inequality, when intersecting with broader social inequality, profoundly affects the lives of young girls from disadvantaged backgrounds. Social disparities rooted in intersecting marginalised identities—such as poverty, race, and ethnicity—intensify the barriers faced by girls who experience multiple forms of exclusion (United Nations Population Fund [UNFPA]

M. van Staden (✉)
Gender Equity Unit, DVC Research and Innovation, University of the Western Cape, Robert Sobukwe Road, Private Bag x 17, Bellville, South Africa
e-mail: mvanstaden@uwc.ac.za

N. V. Roman et al. (eds.), *Social Inequalities Within and Among Countries*, Sustainable Development Goals Series, https://doi.org/10.1007/978-3-032-08741-6_15

and United Nations Children's Fund [UNICEF] 2009). This issue directly relates to Sustainable Development Goal (SDG) 10.2, which calls for the empowerment of all individuals and the promotion of their social, economic, and political inclusion, regardless of age, sex, disability, race, ethnicity, origin, religion, or economic status (United Nations [UN] 2015a). It is also closely connected to SDG 10.3, which seeks to eliminate discriminatory practices and implement appropriate policies to ensure equal opportunities and reduce inequalities of outcome (UN 2015b).

The reality is that girls who exist at the nexus of intersecting marginalised identities—such as being rural, female, from a historically oppressed racial group, and from low-income backgrounds—face compounded oppression and discrimination based on their race, class, and gender. These intersectional disadvantages limit their access to education, health care, economic opportunities, and protection from harm during their youth and as they transition into adulthood. This intersectional oppression exacerbates their plight, as the barriers they encounter in their youth can have lasting effects into adulthood.

Globally, the girl-child not only encounters difficulties in accessing education but also faces gender-specific challenges (Deloumeaux 2019; Evans et al. 2023). The promotion of the right to education for girls extends beyond mere access; it is about ensuring that female learners are safe whilst at school and can equally benefit from the opportunities that education provides (World Bank n.d.). Worker abuse was widespread and included harsh labour practices, physical violence, substandard living conditions, forced evictions, and the notorious *dop system*, in which labourers were partly paid with alcohol. These injustices were further entrenched during the apartheid era, as racially discriminatory laws and unequal access to education sustained a deeply imbalanced labour market. White South Africans retained control over the majority of land and capital, whilst impoverished black South Africans were confined to homelands or relegated to poorly paid farm labour (Devereux 2020). This institutionalised racial discrimination systematised the inhumane treatment of farm workers for centuries (Devereux 2020). Girls growing up in these marginalised farming and rural communities faced layered vulnerabilities linked to economic isolation, discrimination rooted in apartheid legacies, and gender biases stemming from patriarchal norms. The interlocking systems of poverty, racism, and sexism converged to create compounded, intersecting barriers for these girls, based on their identities as female, rural, low-income, and often belonging to historically oppressed racial and ethnic groups in the aftermath of apartheid.

The chapter positions non-formal education programmes as a vital social intervention and a complementary approach to formal schooling, specifically aimed at empowering and providing skills training to adolescent girls who face multiple contextual barriers and inequalities in accessing quality education. It emphasises the importance of understanding and addressing the intersecting inequalities and challenges that disadvantaged groups, particularly girls, encounter in order to ensure they receive equal opportunities for success. Kwauk et al. (2018) support the argument that the non-formal education sector is well positioned to provide skill development opportunities for populations that have been underserved or failed by the formal education system. The chapter argues that non-formal education offers flexibility and innovation often absent in traditional formal education systems, and it challenges the systemic barriers that hinder access to education for marginalised groups such as girls. By exploring a non-formal education programme as a means to equip marginalised girls with skills and pathways to overcome contextual barriers to education, the chapter directly responds to the mandate of SDG indicator 10.3, which seeks to 'ensure equal opportunity and reduce inequalities of outcome' by addressing discriminatory practices that limit girls' access to education.

15.2 Literature Review

15.2.1 Girl-Child's Access to Education

The promotion of gender equality and girls' education has been a prominent feature in the multilateral frameworks of both Education for All (EFA) (United Nations Educational, Scientific and Cultural Organisation [UNESCO] 2000) and the Millennium Development Goals (MDGs) (UN 2000). This remains a key focus in the agenda of the SDGs (UN 2015a). Globally, the number of girls not enrolled in school has reached 62 million, whilst millions struggle to remain in attendance (World Bank 2016). Regrettably, an estimated 100 million girls are expected to withdraw from primary education before completing it (Iwalaiye et al. 2016). The issue is particularly severe for girls in developing countries, who face multifaceted and often dangerous obstacles in their pursuit of education. Even in cases where girls can attend school, they frequently lack the necessary resources and support for learning (Muriungi and Andima 2020).

In Africa, gender discrimination is a pervasive issue that begins early in the lives of girls (Organisation for Economic Co-operation and Development [OECD] 2021). This is due to the deeply ingrained patriarchal systems that have become an integral part of many societies over generations. Such systems perpetuate gender inequality, limiting the opportunities available to girls and women. Consequently, girls face significant obstacles in accessing education, health care, and employment. Addressing gender discrimination and promoting gender equality are critical for the social, economic, and political development of African nations. Akala's (2019) comparative study on women's access to education in Kenya and South Africa highlights that, despite the adoption of progressive gender equality and equity policies in both countries, higher education remains largely inaccessible to many women—particularly those from disadvantaged backgrounds.

The post-apartheid government inherited an education system marked by division and inequality (Mphahlele and Maphalala 2023). More than 25 years into democracy, South Africa still grapples with the enduring legacy of apartheid-era racial segregation and inequalities in its education system. Reforming education to promote equal opportunities for all students challenge. Du Plessis and Mestry (2019) note that educational institutions in rural areas face unique obstacles shaped by their specific context. Barriers to effective education include a lack of parental interest, insufficient state funding, limited resources, underqualified teachers, and multi-grade teaching (Du Plessis and Mestry 2019).

In South Africa, various rural areas and farming communities experience elevated levels of poverty. Consequently, families are incapable of paying for school fees, uniforms, transportation to schools, and other essential educational expenses for their daughters. Patriarchal gender norms remain prevalent in some rural communities, where boys' education is often prioritised over that of girls. Girls are often expected to assist with domestic chores and childcare, which can limit their ability to attend school. In rural areas, schools are frequently scarce and located far from many households. Research by the World Bank (n.d.) on girls' education indicates that those who are forced to walk long distances to school face an elevated risk of violence, with many experiencing violence both on the way to school and within the school environment. Rural schools frequently lack basic infrastructure like electricity, water, sanitation facilities, libraries, and laboratories—factors that undermine the quality of education (Du Plessis and Mestry 2019). Whilst progress has been made, these intersecting challenges help explain why girls from South Africa's rural areas and farming communities continue to face significant educational barriers and lower enrolment, performance, and completion rates compared to their urban and male counterparts.

15.2.2 Girls Living on Farms in the Western Cape, South Africa

Girls living on farms represent a largely understudied population. Existing research has primarily concentrated on two key areas: the conditions of farm workers and the prevention of alcohol and substance abuse. Studies by Devereux (2020), Andrews (2021), and Swanby (2021) have examined the living and working conditions of farm workers. Whilst these contributions are valuable, they often overlook the specific experiences and needs of girls within these communities. Additionally, a substantial body of research has focussed on educational programmes targeting farm communities, particularly those aimed at raising awareness about the risks of alcohol and substance abuse during pregnancy and breastfeeding (Lubbe et al. 2017).

The primary aim of these educational programmes is to prevent foetal alcohol syndrome (FAS) and foetal alcohol spectrum disorders (FASDs). Many girls living on farms in the Western Cape have spent their entire lives in these settings, as their parents are employed as farmworkers. These girls typically attend primary schools that the local farming community. For high school education, they must travel to the nearest towns using transport that is sometimes arranged by farm owners or accessed through subsidised bus services. However, the need to commute limits their ability to participate in extracurricular activities such as sports, drama, or music classes offered at the schools.

According to the Women on Farms Project (WFP) (Devereux 2020), individuals—both male and female—who work and reside on commercial farms in the Western Cape of South Africa face a range of daily challenges. These include, but are not limited to, serious violations of labour rights, poor health conditions, incidents of violence, evictions, and food insecurity (Devereux 2020). Life on farms is marked by harsh living and working conditions that often border on the inhumane.

A parliamentary submission prepared by the WFP further highlights the interconnected socio-economic, health, and educational challenges that significantly affect the lives of female farm workers and dwellers (Parliament Monitoring Group 2022). The research reveals that women employed in the farming sector face particularly arduous challenges. Evidence indicates a persistent trend of intergenerational transmission of socio-economic disadvantage. Consequently, young girls raised in these farming communities are highly likely to follow life trajectories similar to those of their mothers and other female adults engaged in agricultural labour (Fonds voor OntwikkelingsSamenwerking [FOS] 2016).

15.2.3 Non-formal Education for the Girl-Child

Almeida and Morais (2024) describe non-formal education as an alternative to formal education, aimed at ensuring the universal right to education. Non-formal education provides valuable opportunities for both young and middle-aged individuals to develop soft skills, including interpersonal, organisational, and conflict resolution abilities, as well as leadership, intercultural awareness, and practical skills (Kapur 2018). It often serves those on the margins of society and is commonly regarded as a 'second chance', offering learning opportunities that may not have been accessible during formal primary or secondary education (Rodriguez-Kiino 2014:196).

Fe y Alegría, an international movement for popular education and social promotion, has been implementing a non-formal education programme for girls in Guatemala aimed at enhancing personal development and fostering the exercise of their rights and responsibilities (UNESCO 2020). The initiative aspires to build a society grounded in greater equality, peace, and mutual support. As of 2020, it had reached 1546 girls through a range of activities and an extracurricular programme designed for young girls aged 6 to 12 (UNESCO 2020). Two core programmes support the development of girls: *Happy Circles* for ages 6–8 and *Girls' Clubs* for ages 9–12. The former focuses on reading, self-concept, and self-care, whilst the latter emphasises logical thinking, literacy, cognitive development, and artistic expression. Both programmes have contributed to enhanced creativity and literacy, whilst motivating girls to engage more deeply in learning. Complementary initiatives include *Parent Workshops*, which address gender-based violence (GBV) and communication, as well as UNESCO's 2020 Volunteer Programme.

15.3 Methodology

This chapter is based on the Western Cape Department of Agriculture's Rural Female Youth Mentorship Project, which was developed in response to the findings of the Agri-Worker Household Census (AWHHC)—a province-wide survey conducted by the department's Farm Worker Development sub-programme (Shumba 2019). The AWHHC highlighted alarmingly high levels of unemployment, prompting the department to initiate this mentorship project with a specific focus on young rural women (Shumba 2019).

The research adopted an action research approach with the primary goal of positive transformation. It sought to explore the perceived impact of the non-formal education intervention on the lived experiences, behaviours, and relationships of the participating girls, as seen through the perspectives and narratives of parents, guardians, and mentors. Guided by the principles of action research, the study followed a cyclical process of planning, implementing the non-formal education programme, evaluating its outcomes, and using the insights gained to inform subsequent cycles of change (MacKinnon et al. 2019).

Purposive sampling was employed to identify 25 girl participants—70% Afrikaans-speaking and 30% isiXhosa-speaking—living on farms and in rural areas across various municipal districts of the Western Cape, South Africa. In addition, five mentors, all youth over the age of 30 from these districts, were selected to support the participants. Parents and guardians residing on the farms were also included in the study.

The non-formal educational workshops for girls living on farms and in rural communities cover topics such as conflict management, substance abuse awareness, gender-based violence awareness, financial literacy, sexual health, and career guidance. A feminist pedagogical approach was adopted by creating safe spaces for expression, centring the girls' lived realities, and facilitating collaborative learning group discussions.

The interviews took place in the natural settings of the parents', guardians', and mentors' homes on the farms, fostering a comfortable and familiar environment for the interviews. Each interview was held in the participant's preferred language to ensure effective communication and cultural sensitivity.

Data were gathered through various interactive methods employed during the non-formal educational programme. These methods encompassed group discussions, breakout sessions, reflective exercises, and the use of flipcharts, post-it notes, and feedback forms produced by the participants during the activities. Additionally, semi-structured interviews were conducted with parents, guardians, and mentors to obtain further insights.

Thematic analysis is a 'method for identifying, analysing, and interpreting patterns of meaning ("themes") within qualitative data' (Clarke and Braun 2017). This study employed Braun and Clarke's (2006) six-phase approach to thematic analysis, which involved the following steps: familiarisation with the data, generation of initial codes, identification of potential themes, review of themes, definition and naming of themes, and the final phase of report writing.

Prior to the youth camp, initial contact was made with the parents and guardians of the participating girls to communicate the objectives of the programme and to secure their commitment, along with that of the participants. Parents and guardians' consent forms authorising their daughters' participation. During the workshop's contractual phase, participants were thoroughly briefed on the programme's purpose, expectations, and activities. They were then provided with consent forms, which they completed. This critical step ensured that all parties were fully informed of the terms and conditions and felt comfortable proceeding with the programme.

15.4 Findings

Rice et al. (2019) assert that 'Intersectionality involves the study of the ways that race, gender, disability, sexuality, class, age, and other social categories are mutually shaped and interrelated through forces such as colonialism, neoliberalism, geopolitics, and cultural configurations to produce shifting relations of power and oppression'. This intersectional lens informed the design and implementation of this non-formal educational programme for marginalised rural girls, as well as the semi-structured interviews conducted with their parents, guardians, and mentors. The following themes illustrate how these intersecting factors were reflected in the lived experiences and insights shared by participants:

15.4.1 Improved Relationships Between Parents/Guardians and Daughters

This theme focuses on overcoming societal expectations and stereotypes. By acknowledging and challenging these intersecting societal norms, parents, guardians, and daughters can begin to foster more authentic, empowering, and inclusive relationships that respect and celebrate each individual's unique identities and experiences (UNICEF 2023).

Parents and guardians reported observing positive changes in their daughters' behaviour, noting increased confidence. As a result, the relationships between daughters and their mothers have notably improved.

> 'My child has always been very quiet, but I am surprised by how she is now … she never spoke very much, but since she came from the camps, she has a lot to say about things'.

> 'She has also had talks with her classmates at school about 'teen pregnancy'.

> 'My daughter is beginning to be more willing to talk to me. She had always been very 'private', but after she came back from the camp, a change is noticeable'.

In both cases, the mothers sound pleasantly surprised and perhaps proud of their daughters' new openness and engagement. The non-formal educational experience appears to have provided a meaningful opportunity for the girls to find their voices and feel more comfortable asserting themselves. Their increased communication, both within the family and with peers, indicates heightened confidence and personal development. The improved relationships and open communication between daughters and parents/guardians also demonstrate how the residential non-formal educational programme promoted the social inclusion of these marginalised girls within their family units. By giving them a voice and a platform for expression, it challenged the norms that previously constrained their inclusion.

15.4.2 Mentoring Support

In this study, mentoring support complemented the efforts of parents and guardians in promoting gender-transformative parenting practices. Daughters received additional guidance from their peers, enabling them to explore their identities, build resilience, and navigate intersectional challenges (Donville 2020). However, the harsh living conditions on the farms negatively affected participants' well-being, making it difficult for them to fully embrace the opportunity. As one mentor noted, the family environment often lacked the capacity to nurture and support the learning process:

> 'Sarah is very introverted to the extent that she hardly makes eye contact when one engages with her. She has been in the care of her grandparents since her mother died a few years ago. The challenge in this household is that the grandparents are in their advanced years and as a result, Sarah is very much left to care for herself'.

> 'This is not to say that her grandparents do not love her, but from what I have observed, they are not really in a position to provide Sarah with the support that she needs to grow into a fully functional and self-sufficient young person. Her siblings are there to support her, but I sense that Sarah needs more than what her nucleus home family'.

The passage highlights Sarah's potential for growth and development through mentorship and guidance that extend beyond the support available within her family. The mentor suggested that participation in the mentorship programme offers Sarah access to new experiences and perspectives, enabling her to develop into a well-rounded individual equipped to navigate life's challenges. The theme of mentoring support underscores how the non-formal education programme addresses unequal opportunities by supplementing familial limitations with additional enrichment provided by mentors. In doing so, it offers marginalised girls more equitable and comprehensive access for to support systems.

15.4.3 Transference of Knowledge and Skills

The act of sharing knowledge and skills is essential to sustaining the impact of learning circles beyond the formal conclusion of the programme. When participants effectively transmit what they have learned to family members and peers, the knowledge becomes embedded within the community. This significantly enhances the potential for long-term behavioural change and the reinforcement of positive outcomes.

> 'Mary has started to teach her younger sister the things she had learnt on the camp by using the training manual she received at the Youth Camp (Mother)'.

> 'Maureen said that she has started to share what she has learnt with her close circle of friends at school. I encouraged her to share what she has learnt with other learners who are not necessarily a part of her close circle of friends (Mentor)'.

These narratives reflect the participants' proactive engagement in sharing their learning—an intended and valued outcome of the programme. Marginalised girls living on farms and in rural communities become agents of knowledge transfer, challenging the intersecting power imbalances that contribute to their marginalisation and silence their voices and talents.

15.5 Discussion

The non-formal educational programme for marginalised adolescent girls in rural farming communities had several positive influences, presenting an intersectional approach aligned with promoting gender equality and inclusion as outlined in SDG 10.2 and SDG 10.3. The social intervention enabled girls to develop greater confidence, communication skills, and openness, which improved their relationships with their parents and guardians. In addition, it provided them with a platform to express themselves, fostering their social inclusion (SDG 10.2) by enhancing their ability to assert themselves and engage in dialogues around topics such as teen pregnancy. The findings illustrated how household environments alone may be insufficient for nurturing the development of marginalised girls, highlighting the need for mentoring support as a measure to provide equal opportunities (SDG 10.3).

Engaging their families and friend circles enabled a ripple of intersectional empowerment to extend beyond just the individual girls in the programme. Building the capacities of local leaders (mentors and girl participants) empowers communities to advocate against discriminatory practices (SDG 10.3). It positions them as owners of information, which could spark wider community dialogue and attitude shifts. This positioning addresses power imbalances and systemic inequalities by centring the perspectives, experiences, and solutions envisioned by the marginalised themselves. It levels an unequal playing field where their knowledge has been historically subjugated (SDG 10.3). Ongoing initiatives play a critical role in sustaining opportunities for underserved groups, particularly girls living on farms and in rural communities, thereby advancing the aims of SDG 10.3.

15.6 Recommendations

Based on the positive findings from the study of the non-formal education programme, the following recommendations are proposed:

- Design a comprehensive strategy to expand the programme's reach and amplify its transformative impact. This may involve collaboration with key stakeholders—such as government agencies, educational institutions, or community organisations—to replicate the programme in different regions or adapt it to diverse contexts.
- Utilise qualitative evidence to advocate for the integration of effective non-formal education practices into formal educational policies and curricula. This could include engaging with policymakers, educational institutions, and relevant stakeholders to promote the adoption of inclusive, participatory, and culturally responsive approaches.
- Enhance mentor and facilitator training with a special focus on pedagogical techniques, inclusive practices, and strategies for effectively engaging with participants from diverse and marginalised backgrounds.

15.7 Conclusion

The non-formal education programme presents a viable option for facilitating gender equality and uplifting girls in underprivileged conditions. It is an attractive alternative to mainstream education with its limitations (Dumitru 2018). By supplementing formal education, non-formal education can provide a more comprehensive learning experience that empowers marginalised communities.

An intersectional framework informs the non-formal educational programme, which has demonstrated a promising influence; however, the findings also revealed additional opportunities to strengthen such social interventions for marginalised girls living on farms and in rural communities. The critical need for mentoring systems beyond just family units was clear. To strengthen this programme, the establishment of community-based mentorship networks that can provide more holistic support and guidance for participants' overall development is essential. This work can also contribute to the body of knowledge on intersectional non-formal educational programme models that can advance understandings of transformative praxis for girls facing compounded inequities.

References

Akala B (2019) Gender inequality in education is still an issue in Kenya and South Africa. https://theconversation.com/gender-inequality-in-education-is-still-an-issue-in-kenya-and-south-africa-125747. Accessed 19 Sept 2024

Almeida F, Morais J (2024) Non-formal education as a response to social problems in developing countries. E-Learn Digit Media 22(2):1–17

Andrews M (2021) Trapped in debt: gendered life and Western cape farms. Gend Quest 9(1):1–10

Braun V, Clarke V (2006) Using thematic analysis in psychology. Qual Res Psychol 3(2):77–101

Clarke V, Braun V (2017) Themat Anal J Posit Psychol 12(3):297–298. https://doi.org/10.1080/17439760.2016.1262613

Deloumeaux L (2019) New methodology shows 258 million children, adolescents and youth are out of school. UNESCO, Paris

Devereux S (2020) Violations of farm workers' labour rights in post-apartheid South Africa. Dev South Afr 37(3):382–404

Donville J (2020) Guidance note: gender transformative education and programming. https://plan-international.org/eu/news/2020/08/20/gender-transformative-education-guidance. Accessed 18 Mar 2025

Dumitru TC (2018) Impact of non-formal education on the efficacy of school learning. Stud Univ Mold Seria Ştiinţe ale Educaţiei 119(9):229–233

Du Plessis P, Mestry R (2019) Teachers for rural schools–a challenge for South Africa. S Afr J Educ 39

Evans DK, Mendez Acosta A, Yuan F (2023) Girls' education at scale. World Bank Res Obs 38(1):47–74. https://doi.org/10.1093/wbro/lkad002

Fonds voor OntwikkelingsSamenwerking (FOS) (2016) Documentary on Women on Farms: "Everyone Together". https://www.fos.ngo/womenonfarms/. Accessed 12 June 2024

Iwalaiye EM, Abah MJ, Johnson UE, Iwalaiye CY, Ali AY (2016) The challenges of girl-child education in government secondary schools of Abuja, Nigeria—implication for counselling. Int J Ser Multidiscip Res 2(4):26–37

Kapur R (2018) Non-formal education. University of Delhi, Delhi

Kwauk C, Braga A, Kim H et al (2018) Non-formal girls' life skills programming: implications for policy and practice. The Brookings Institution, Washington, DC

Lubbe M, Van Walbeek C, Vellios N (2017) The prevalence of foetal alcohol syndrome and its impact on a child's classroom performance: a case study of a rural south African school. Int J Environ Res Public Health 14(8):896

MacKinnon GR, Young D, Paish S et al (2019) Preparing instructional leaders: evaluating a regional program to gauge perceived effectiveness. Int J Educ Policy Leadersh 14(1):1–20

Mphahlele RS, Maphalala MC (2023) Contextualising rural education in south African schools, vol 3. Brill, Leiden. https://doi.org/10.1163/9789004547025_001

Muriungi C, Andima G (2020) A gender community outreach for girls living in vulnerable contexts: experiences from school literacy clubs in Tharaka Nithi County, Kenya. https://code.ngo/wp-content/uploads/2020/05/A-Gender-Community-Outreach-for-Girls-Living-in-Vulnerable-Contexts-Kenya-May-2020.pdf. Accessed 24 Mar 2025

Organisation for Economic Co-operation and Development (OECD) (2021) SIGI 2021 regional report for Africa: social institutions and gender index. OECD Publishing, Paris

Parliamentary Monitoring Group (2022) Women on Farms Project (WFP). https://static.pmg.org.za/220316WFPSubmission_AppropriationsCommittee_11March22_11.pdf. Accessed 16 Sept 2024

Rice C, Harrison E, Friedman M (2019) Doing justice to intersectionality in research. Cult Stud Crit Methodol 19(6):409–420. https://doi.org/10.1177/1532708619829779

Rodriguez-Kiino D (2014) Reaching the margins: critical feminist practices in nonformal education for street children. Int J Crit Pedagog 5(2):194–210

Shumba B (2019) Making an impact on young women in rural areas. Agri 16(1):37–39

Swanby H (2021) Embracing humanity and reclaiming nature: neglected gendered discourses in the struggles of Western cape farm workers. Gend Quest 9(1):1–14

United Nations (UN) (2000) United Nations millennium declaration: resolution adopted by the general assembly. 55th session, 18 September. United Nations, New York

United Nations (UN) (2015a) The UN sustainable development goals. United Nations, New York. http://www.un.org/sustainabledevelopment/summit. Accessed 18 Mar 2025

United Nations (UN) (2015b) Transforming our world: the 2030 agenda for sustainable development: resolution adopted by the general assembly on 25 September 2015. United Nations, New York

United Nations Children's Fund (UNICEF) (2023) Supporting families for gender transformative parenting. Challenging the gender norms and gender inequalities to create a more equitable environment for all children to thrive. UNICEF, New York. https://www.unicef.org/reports/resource-gender-transformative-parenting

United Nations Educational, Scientific and Cultural Organisation (UNESCO) (2000) The Dakar framework for action: education for all: meeting our collective commitments. UNESCO, Paris

United Nations Educational, Scientific and Cultural Organisation (UNESCO) (2020) Non-formal education, sustainable development and the education 2030 agenda. UNESCO, Paris

United Nations Population Fund (UNFPA) and United Nations Children's Fund (UNICEF) (2009) Girl power and potential: a joint programming framework for fulfilling the rights of marginalized adolescent girls. https://www.unfpa.org/publications/girl-power-and-potential. Accessed 25 Nov 2023

World Bank (n.d.) Girls' education. https://www.worldbank.org/en/topic/girlseducation. Accessed 8 Dec 2023

World Bank (2016) Education global practice: reaching girls, transforming lives. World Bank Publishing, Washington, DC

Part III
Evaluating the SDGs

Understanding the Evaluation System for the Sustainable Development Goals (SDG): Inequalities Amongst Countries

16

R. Ramli and N. I. Z. Yahaya

Abstract

The Global Agenda 2030, widely known as the Sustainable Development Goals (SDGs), was adopted by United Nations (UN) member states in 2015 as a universal call to action to address the world's most pressing challenges by 2030. Effective monitoring and evaluation (M&E) are essential for tracking progress, ensuring accountability, and informing policy decisions. This chapter examines the UN's M&E system for the SDGs, with a focus on its implementation at the global, regional, and national levels. At the global level, the United Nations Statistical Commission and the Inter-agency and Expert Group on SDG Indicators (IAEG-SDGs) play key roles in data collection and reporting. Regionally, organisations such as ESCAP support data harmonisation and regional assessments. At the national level, countries incorporate SDG indicators into their development plans, working through national statistical offices and collaborating with civil society. Voluntary National Reviews (VNRs) provide an important platform for member states to report on their progress and share accomplishments. The discussion underscores the challenges and opportunities in achieving comprehensive, accurate, and timely evaluations. Notably, the Mid-Term Review in 2023 revealed a concerning rise in inequality amongst countries. By analysing the UN's multi-tiered M&E framework alongside the Impact Ranking System, this chapter outlines best practices and offers insights into future directions for understanding and addressing inequality.

Keywords

Monitoring · Evaluation · Asia-Pacific · Inequalities · Development

16.1 Introduction

As World War II drew to a close in 1945, much of the world lay in devastation, and the global yearning for peace was profound. From 25 April to 26 June 1945, representatives from 50 countries gathered at the United Nations Conference on International Organisation in San Francisco, California. Over the course of two months, they drafted and ultimately signed the United Nations (UN) Charter, establishing a new international body—the United Nations—with the goal of preventing another catastrophic global conflict (UN 2024a, b). The UN pursues its mission by working to prevent conflicts, assisting parties in

R. Ramli (✉)
Universiti Poly-Tech Malaysia (UPTM), Kuala Lumpur, Malaysia
e-mail: rashilaramli@uptm.edu.my

N. I. Z. Yahaya
Institute of Malaysian and International Studies (IKMAS), UKM, Bangi, Selangor, Malaysia
e-mail: p127338@siswa.ukm.edu.my

N. V. Roman et al. (eds.), *Social Inequalities Within and Among Countries*, Sustainable Development Goals Series, https://doi.org/10.1007/978-3-032-08741-6_16

conflict to make peace, deploying peacekeepers, and fostering conditions that enable peace to be sustained and thrive.

The development of countries always remains at the centre of UN interest. The Millennium Development Goals (MDGs) served as the development framework from 2001 to 2015. Whilst the MDGs achieved success, particularly amongst African states, a more sustainable development plan was necessary to reduce inequality gaps amongst countries. In a similar spirit of collaboration, the UN (2018) reported that 193 member states, along with numerous civil society groups, academic institutions, and private-sector stakeholders, engaged in an open, democratic, and participatory negotiation process. This effort culminated in the proclamation of the 2030 Agenda for Sustainable Development, featuring its Sustainable Development Goals (SDGs), in September 2015. The 17 goals are further delineated into 169 targets, creating a list that is criticised for being both overly extensive and lacking sufficient detail for each individual item (Easterly 2015). This led to the establishment of the role of Monitoring and Evaluation (M&E) of the UN's 2030 SDGs, which is a management process that integrates oversight (monitoring) with the evaluation of choices, processes, decisions, actions, and outcomes (evaluation) (Persaud and Dagher 2021).

Any development programme must have an effective monitoring and evaluation programme to ensure that quality implementation meets the intended target. This chapter surveys different M&E mechanisms at the global, regional, and national levels. Indirectly, it provides insight into understanding inequalities amongst countries.

16.2 Understanding the Role of M&E in the SDGs Framework

The discussion on the role of M&E in this section centres on three key issues: ensuring accountability and transparency in the implementation of the SDGs; providing data and insights for evidence-based policymaking and resource allocation; and identifying successes, challenges, and areas requiring improvement to achieve the SDGs.

Accountability and transparency are essential for the successful implementation of the SDGs. Accountability ensures that stakeholders are held responsible for their actions and decisions in pursuit of the SDGs, whilst transparency ensures that these actions and decisions are visible and accessible to all relevant parties (SDG Accountability Handbook 2024). In the context of SDG implementation, accountability involves holding governments, international organisations, civil society, and other actors answerable for their commitments to the goals (Breuer et al. 2021; Transparency, Accountability and Participation [TAP] Network 2024). This includes monitoring progress, evaluating impact, and openly addressing challenges. Transparency, meanwhile, involves making information about SDG activities, resources, and outcomes readily available to the public, enabling informed decision-making and fostering trust amongst stakeholders (Glass and Newig 2019). By prioritising accountability and transparency, the SDG framework can promote greater participation, collaboration, and collective action towards sustainable development for all.

M&E also plays a pivotal role in providing data and insights for evidence-based policymaking and resource allocation (United Nations Children's Fund [UNICEF] 2008). For example, consider a country aiming to address poverty eradication as part of its SDG commitments. Through robust M&E mechanisms, policymakers can continuously track progress on poverty reduction initiatives, analysing data on income levels, access to education, health care, and basic services. By collecting and interpreting these data, they can identify which interventions are yielding positive results and which are falling short (Kusek and Rist 2004). They may, for instance, discover that investments in education and vocational training have significantly increased employment rates in low-income communities. Armed with such evidence, policymakers can make informed decisions on where to allocate resources—scaling up successful programmes and redirecting funds away from less effective ones. Furthermore, real-time feedback allows them to refine policies and strategies, ensuring that efforts remain targeted, efficient, and impactful. In this way, M&E serves as a cornerstone for evidence-based policy-making, enabling governments to take informed actions that drive sustainable development and improve the lives of their citizens.

Finally, identifying successes, challenges, and areas needing improvement is essential for ensuring meaningful progress towards sustainable development. Successes signal positive outcomes from policies, interventions, or initiatives aligned with the SDGs (UN 2022), highlighting achievements such as poverty reduction, improved access to education and health care, or advancements in environmental conservation. However, we must also acknowledge the persistent challenges that hinder progress. These may include inadequate funding, political instability, weak infrastructure, social inequalities, or the impacts of climate change (Dugarova and Gülasan 2017). Recognising such challenges enables stakeholders to devise targeted strategies to overcome obstacles and advance SDG targets. Moreover, identifying areas requiring improvement is critical for refining approaches and enhancing the overall effectiveness of SDG implementation (Nonet et al. 2022). Such areas may involve gaps in data collection and analysis, limited policy coherence, institutional capacity deficits, or insufficient inclusivity in decision-making processes. By pinpointing these, stakeholders can prioritise actions, allocate resources strategically, and foster collaboration to accelerate progress towards the SDGs (United Nations Development Programme [UNDP] 2016). A thorough assessment of successes, challenges, and improvement needs therefore offers a roadmap for navigating the complexities of the SDG agenda and advancing global sustainable development.

16.3 SDG Global Level M&E

The UN has developed a robust framework for monitoring and evaluating progress towards the SDGs at the global level. This framework is built on the Global Indicator Framework, which includes 231 unique indicators designed to comprehensively track the 17 goals and 169 targets. It ensures standardisation, enabling consistent and comparable data collection across countries (UN 2024a, b).

Key actors in global M&E include the following:

- **United Nations Statistical Commission:** This body is responsible for overseeing the development and implementation of the global indicator framework, ensuring that the indicators are scientifically sound and statistically feasible.
- **Inter-agency and Expert Group on SDG Indicators (IAEG-SDGs):** Composed of member states and international agencies, the IAEG-SDGs plays a crucial role in refining indicators, addressing methodological issues, and coordinating data collection efforts.
- **United Nations Department of Economic and Social Affairs (UNDESA):** This agency provides critical support for global M&E efforts, including data analysis, capacity-building initiatives, and technical assistance to member states.

National statistical offices provide data for the global M&E process, which is then compiled into global reports.

Key mechanisms include the following:

- **Annual SDG Progress Reports**: Published by the UN, these reports provide a comprehensive overview of global progress towards the SDGs. They highlight achievements, identify emerging trends, and pinpoint areas needing additional attention.
- **Voluntary National Reviews (VNRs):** Presented by countries at the High-Level Political Forum (HLPF), VNRs offer a platform for nations to share their experiences, successes, and challenges in implementing the SDGs, fostering mutual learning and accountability.

Global M&E faces challenges such as data gaps, particularly in developing countries, which hinder the accurate tracking of goals like SDG 1 (No Poverty) and SDG 3 (Good Health and Well-being). Inconsistencies in data quality also affect the measurement of progress on goals like SDG 4 (Quality Education) and SDG 6 (Clean Water and Sanitation). However, advancements in technology, such as big data and remote sensing, offer opportunities to enhance data collection and analysis. These innovations can improve the monitoring of goals such as SDG 11 (Sustainable Cities and Communities) and SDG 13 (Climate Action), making tracking more accurate and timely. The figure below indicates that, except for SDG 2 (Hunger), SDG 11 (Sustainable Cities and Communities), and SDG 14 (Life Below Water), all other goals are making fair progress in the implementation of the SDGs. However, it should also be noted that levels of stagnation and regression on SDG 2 (Hunger), SDG 8 (Decent Work), and SDG 14 (Life Below Water) are also prevalent. This is apparent in Fig. 16.1.

16.4 SDG Regional Level M&E

Regional organisations play a crucial role in harmonising M&E efforts across countries, addressing region-specific challenges, and fostering cooperation amongst member states. They provide platforms for sharing best practices, lessons learned, and resources.

Key regional actors include the following:

- **Regional Bodies:** Organisations such as the African Union (AU), European Union (EU), and Association of Southeast Asian Nations (ASEAN) lead regional initiatives on the SDGs. They coordinate efforts amongst member states and provide strategic guidance.
- **UN Regional Commissions:** Commissions such as the United Nations Economic and Social Commission for Asia and the Pacific (UNESCAP), the United Nations Economic Commission for Africa (UNECA), and the United Nations Economic Commission for Latin America and the Caribbean (UNECLAC) support regional M&E through

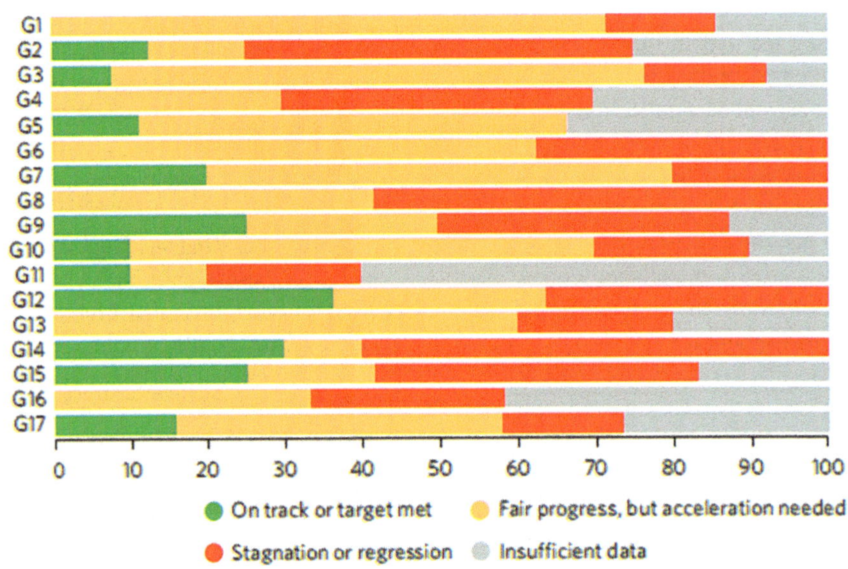

Fig. 16.1 In the SDG mid-term review 2023, Bar chart showing progress across 17 goals (G1 to G17) with color-coded segments: green for "On track or target met," yellow for "Fair progress, but acceleration needed," red for "Stagnation or regression," and gray for "Insufficient data." Each goal has varying proportions of these categories, indicating different levels of achievement and data availability. Source: United Nations Statistics Division (UNSD 2025)

data collection, analysis, and policy advice. They help to harmonise data standards and methodologies across countries within their regions.

Regional M&E processes focus on data harmonisation, capacity-building, and reporting. Key mechanisms include **Regional SDG Reports,** which provide a comprehensive overview of progress, challenges, and opportunities at the regional level. They help to identify regional trends and areas that require collective action. **Regional Forums and Commissions** are platforms that facilitate dialogue, coordination, and the sharing of best practices amongst countries. They promote regional cooperation and help to align national efforts with regional priorities.

Regional M&E faces challenges such as data disparities, varying capacities amongst countries, and regional conflicts. However, opportunities exist to enhance regional data infrastructure, promote inter-regional learning, and leverage regional cooperation to address common challenges. Strengthening regional partnerships and integrating regional priorities into global and national frameworks can also enhance M&E effectiveness. Regional M&E faces challenges such as data disparities and varying capacities amongst countries, which can impede the tracking of goals like SDG 2 (Zero Hunger) and SDG 5 (Gender Equality). Regional conflicts can also disrupt efforts towards SDG 16 (Peace, Justice, and Strong Institutions). However, opportunities exist to enhance regional data infrastructure, which can improve monitoring for SDG 9 (Industry, Innovation, and Infrastructure) and SDG 7 (Affordable and Clean Energy). Promoting inter-regional learning can facilitate the exchange of best practices for achieving SDG 12 (Responsible Consumption and Production) and SDG 14 (Life Below Water). Strengthening regional partnerships can also advance progress on SDG 13 (Climate Action) and SDG 15 (Life on Land).

Figure 16.2 presents the SDG progress data for the Asia-Pacific region. As shown, none of the SDGs are currently on track to meet the mid-term review target, and one SDG has even regressed in terms of its target and implementation. It is particularly concerning to note the lack of sufficient indicators in the available data. However, when compared to global data, the Asia-Pacific region appears to be making fair progress on most goals, with the notable exception of SDG 13 (Climate Action) and SDG 14 (Life Below Water).

Another global/regional mechanism is the Sustainability Self-Evaluation Tool for Higher Education Institutions, or SET4HEI, launched by UNESCO on 15 July 2024. This instrument is a combined effort of three institutions: the UNESCO International Institute for Higher Education in Latin America and the Caribbean, the United Nations Academic Impact Initiatives, and the UN Higher Education Sustainability Initiative (HESI).

There are various perspectives in the process of tracking universities' SDG performance. Some focus on risk analysis in measuring climate action, the need for an interdisciplinary approach, and capacity building for sustainability practices. SET4HEI is a tool that allows higher education institutions (HEIs) to go online and answer 18 self-assessment questionnaires. Upon completion of the self-assessment across all 17 goals or a selection of goals, participants receive a score indicating their level of sustainability practices. They can then pursue certification through SET4HEI.

Another HEI evaluation tool for universities is the Times Higher Education (2024) Impact Ranking. Under this system, four broad areas for assessment are research, stewardship, outreach, and teaching. Universities must provide data on SDG 17 (22%) and three other SDGs at 26% each, making a total of 100%. Amongst the four partner universities in the IWAAC-10 Project, only the University Kebangsaan Malaysia (UKM) participated in the Impact Ranking. Taking the best university from all four countries, Table 16.1 shows the ranking of the top universities from the four partner countries.

As shown in the table, within SDG 17, which focuses on partnerships, all scores are 80 and above. Targets such as Research into Partnerships for the Goals (17.1), Relationships to Support the Goals (17.2), Publications of SDG Reports (17.3), and Education for the SDGs (17.4) are areas that all countries concentrated on. However,

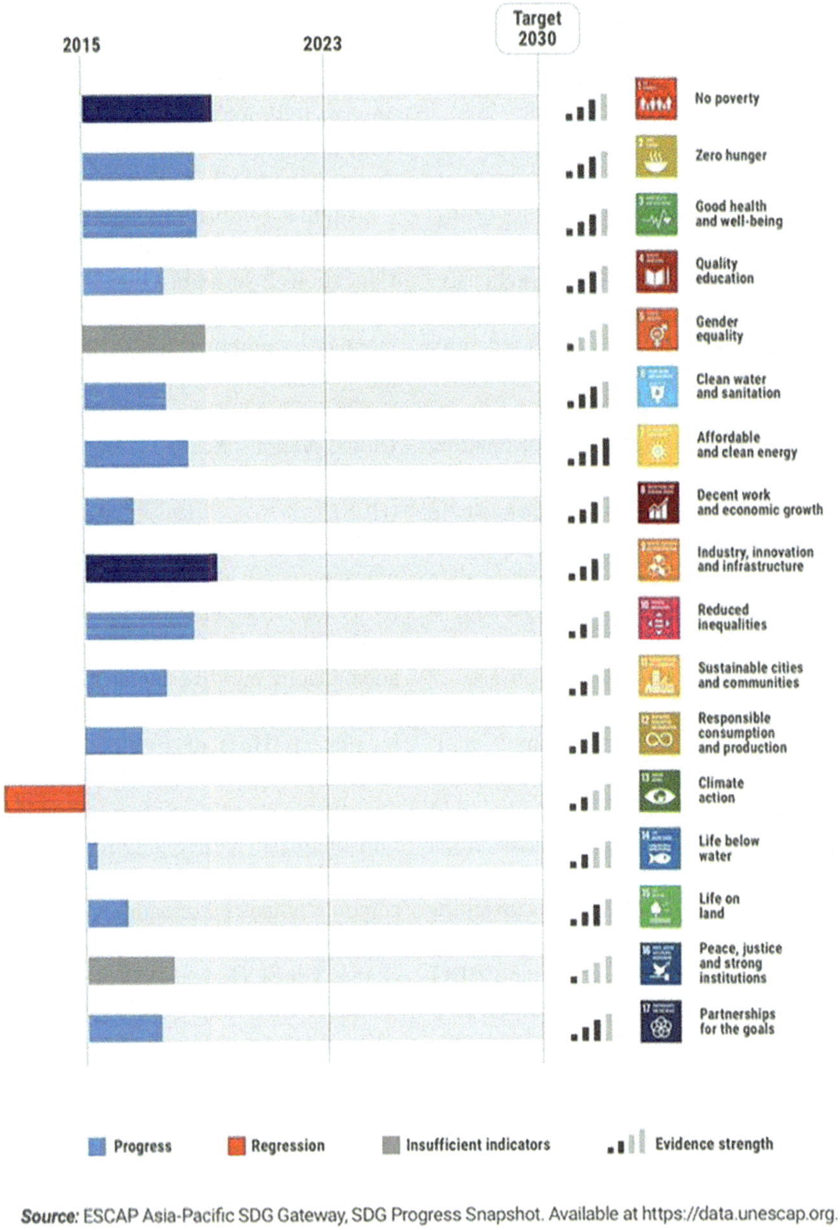

Fig. 16.2 SDG progress implementation in Asia-Pacific

there is a variation in the choices of the other three SDGs. The most common goal in three countries, excluding Malaysia, is SDG 5 on Gender Equality, where the score ranged from 71 to 86 points. Malaysia may not have selected SDG 5 because there are overwhelmingly more female than male students on most Malaysian campuses. Additionally, SDG 10 on Reducing Inequalities was not listed at all in the four countries.

Table 16.1 SDG impact ranking 2024: Four-country comparison

Country	Best University					Score	Rank
Malaysia	Universiti Sains Malaysia	G16	G2	G6	G17		18
		90.6	80.3	81.3	97.4	95.0	
South Africa	University of Johannesburg	G1	G5	G8	G17		36
		88.8	75.9	84.7	84.7	93.0	
India	Amrita Vishva Vidyupekam	G4	G3	G5	G17		81
		90.5	91.3	74.1	80	90.3	
Germany	Free University of Berlin	G5	G16	G12	G17		101-200
		71.1	80.3	81.4	84.7	84-89	

Source Impact Ranking 2024 (Online)

16.5 SDG National Level M&E

At the national level, countries are responsible for integrating the SDGs into their development plans and policies. This involves developing national indicator frameworks tailored to local contexts and priorities, ensuring that SDG implementation is relevant and effective. Key national actors include the following:

- **National Governments and Statistical Offices:** These entities lead efforts in collecting, analysing, and reporting data on SDG indicators, playing a pivotal role in ensuring that the data are accurate and reliable.
- **Civil Society and the Private Sector:** Civil society organisations and the private sector contribute to the M&E process by providing additional data, monitoring progress, and advocating for effective policies and practices.

Processes and mechanisms include national M&E systems that involve regular data collection through various means, such as censuses, surveys, and administrative records. Key processes include the following:

- **National SDG Reports:** These reports detail a country's progress and challenges in implementing the SDGs. They serve as a basis for policy adjustments and strategic planning.
- **VNRs:** VNRs are important reports that provide insights into countries' progress on the SDGs. They promote international accountability and encourage countries to share best practices and learn from each other. The four countries participating in the IWAAC-10 project have presented their VNRs at the UN High-Level Political Forum (HLPF). Table 16.2 indicates the presentations of the VNR by the four countries.

Table 16.2 Country VNR presentations at HLPF

Countries	Year VNR delivered	Next VNR
Germany	2016, 2021	2025
India	2017, 2020	2025
Malaysia	2017, 2021	2025
South Africa	2019	2025

National M&E systems face challenges such as limited technical capacity, inadequate funding, and political instability. However, there are opportunities to leverage local knowledge, innovative practices, and technological advancements to improve data accuracy and relevance. Enhancing collaboration between government entities, civil society, and the private sector can also strengthen national M&E efforts. These challenges impact the monitoring of goals like SDG 8 (Decent Work and Economic Growth) and SDG 9 (Industry, Innovation, and Infrastructure).

Political instability can disrupt data collection for goals like SDG 16 (Peace, Justice, and Strong Institutions). Nevertheless, opportunities exist to leverage local knowledge and innovative

Table 16.3 Classification of projects (EC/SC/EV)

No	Region	EC	SC	EV
1	Northern 1	23	22	11
2	Northern 2	19	9	4
3	Central 1	10	12	7
4	Selangor	18	10	4
5	Eastern	16	15	3
6	South-Johor and Pahang	8 7	4 4	– 1
7	Sabah east	5	7	4
8	Sabah west	16	14	7
9	Sarawak north	16	11	1
10	Sarawak south	19	9	2
	Total	157	101	44

Legend EC: Economics, SC: Social, EV: Environment.
Source Ramli et al. (2024)

practices to improve data accuracy and relevance. For instance, community-driven data collection can enhance the monitoring of SDG 10 (Reduced Inequalities) and SDG 11 (Sustainable Cities and Communities). Strengthening collaboration between government entities, civil society, and the private sector can also bolster efforts towards SDG 17 (Partnerships for the Goals).

Finally, as an example of localising SDGs from a national level to the local level, Malaysia worked through the Malaysian Parliament to implement SDG projects. By 2024, 115 out of 222 parliamentary constituencies will have worked to localise SDG projects. When impact evaluations were conducted on projects between 2020 and 2022, it was identified that projects at parliamentary levels tended to focus on economic or social issues, with the environment receiving less attention. As shown in Table 16.3, out of all projects evaluated in 2023, only 44 were predominantly focussed on the environment.

By examining the data above, it was found that issues related to increasing income generation, upskilling, and utilising digital tools resonate more with people than environmental concerns. The impact evaluation also highlights the importance of building networks and learning new skills.

16.6 Conclusion and Recommendations

This chapter has presented examples of global M&E through VNRs conducted by countries at the UN. It also highlighted M&E tools used by HEIs, such as the Impact Ranking on SDGs, as another example of SDG implementation within academia. This multilevel approach enables a comprehensive understanding of progress and facilitates coordinated efforts to address global, national, and regional development challenges. Notably, in mapping the IWAAC-10 four-country study, Gender Equality was featured in the Impact Rankings of three of the participating countries.

A robust M&E system is critical for tracking progress towards the SDGs at all levels—global, national, regional, and local. At the global level, the VNR consultation process should actively engage those furthest from the centre, including indigenous peoples, refugees, and stateless individuals. At the national level, mechanisms to localise the SDGs at the grassroots should be established to ensure more meaningful and substantive impacts. At the individual level, people's actions must align with the principles of sustainability and inclusivity.

Whilst challenges such as data gaps and capacity constraints remain, there are abundant opportunities to strengthen M&E through technological innovations, partnerships, and creative practices. By cultivating a culture of accountability, learning, and continuous improvement, stakeholders can help ensure that M&E systems effectively support the achievement of the SDGs.

References

Breuer A, Leininger J, Brosbøl K et al (2021) Key players in national SDG accountability: the role of parliaments. German Development Institute, Bonn. https://doi.org/10.23661/bp15.2021.v1.1

Dugarova E, Gülasan N (2017) Global trends: challenges and opportunities in the implementation of the sustainable development goals. United Nations Development Programme, New York

Easterly W (2015) The trouble with sustainable development goals. Current History 114(775):322–324

Glass L, Newig J (2019) Governance for achieving the sustainable development goals: how important are participation, policy coherence, reflexivity, adaptation and democratic institutions? Earth Syst Gov 2:100031. https://doi.org/10.1016/j.esg.2019.100031

Times Higher Education (2024) University Impact Rankings 2024. https://www.timeshighereducation.com/impactrankings. Accessed 9 May 2025

Kusek JZ, Rist RC (2004) Ten steps to a results-based monitoring and evaluation system. World Bank, Washington DC. https://doi.org/10.1596/0-8213-5823-5

Nonet GA, Gössling T, van Tulder R, Bryson JM (2022) Multi-stakeholder engagement for the sustainable development goals: introduction to the special issue. J Bus Ethics 180(4):945–957. https://doi.org/10.1007/s10551-022-05192-0

Persaud N, Dagher R (2021) The role of monitoring and evaluation in the UN 2030 SDGs agenda. Springer, Cham. https://doi.org/10.1007/978-3-030-70213-7

Ramli R, Daud S, Sharifah Syahirah SS et al (2024) Hard facts on localising SDGs in Malaysia. PSSM, Kuala Lumpur

SDG Accountability Handbook (2024) What does the 2030 Agenda say about accountability? https://www.sdgaccountability.org/background/. Accessed 9 May 2025

Transparency, Accountability & Participation (TAP) Network (2024) Civil society spotlight reports. https://www.sdgaccountability.org/. Accessed 9 May 2025

United Nations (UN) (2018) The 2030 agenda and the sustainable development goals: an opportunity for Latin America and the Caribbean. United Nations, Santiago

United Nations (UN) (2022) The sustainable development report 2022. United Nations, New York

United Nations (UN) (2024a) History of the United Nations. https://www.un.org/en/about-us/history-of-the-un. Accessed 9 May 2025

United Nations (UN) (2024b) Maintain international peace and security. https://www.un.org/en/our-work/maintain-international-peace-and-security/. Accessed 9 May 2025

United Nations Children's Fund (UNICEF) (2008) Bridging the gap: the role of monitoring and evaluation in evidence-based policy making. UNICEF, New York

United Nations Development Programme (UNDP) (2016) Transitioning from MDGs to the SDGs. United Nations Development Programme, New York

United Nations Statistics Division (UNSD) (2025) II. Sounding the alarm: SDG Progress at the midpoint. https://unstats.un.org/sdgs/report/2023/progress-midpoint/. Accessed 9 May 2025

From the Local to Global: Impact Evaluation on SDG Solution Projects in Malaysia

17

Raan Hann Tan, Yuen Beng Lee, Nurul Hidayat Ab Rahman, and Maslisa Zainuddin

Abstract

According to the UN Secretary-General, SDG 10 remains the least-performing SDG. Action on this goal has therefore never been more urgent. Addressing this urgency is critical, as continued inequality causes global unemployment, health, and environmental issues. In Malaysia, there is increasing concern over the sustainable livelihood, economic growth, and social stability of those working in informal job sectors, making them vulnerable to increased social inequalities. Such forms of inequality create disparities in the distribution of wealth, access to opportunities, and potential societal biases. In 2020, the All-Party Parliamentary Group Malaysia on the Sustainable Development Goals (APPGM-SDG) launched solution projects to improve the livelihood of the B40 communities. Impact evaluations are vital to assess the effectiveness and impact of interventions. In 2022, an evaluation team examined the implications of these solution projects across parliamentary constituencies in the Central Malaysian region of Selangor, Kuala Lumpur, and Pahang. Using document analysis, focus group discussions, and site visits, the team identified the impact evaluation using the deep, clear, wide, high, SDGs, and gender domains. A report on the impact evaluation of SDG efforts, psychological effects, and the efficiency of the projects was written. Using the SDG 10: Reduced Inequalities framework and Theory of Change, this chapter discusses how social interventions reduce income inequalities, their impacts, and the challenges beneficiaries face.

Keywords

APPGM-SDG Malaysia · Theory of change · B40 community

17.1 Introduction

The UN Secretary-General has identified SDG 10 (Reduced Inequalities) as the least-performing goal globally (PATHFINDERS 2023). Malaysia similarly faces challenges in achieving SDG 10, with persistent inequalities reflected in the high Gini coefficients recorded in 2015 and 2018, particularly affecting vulnerable groups such as women, minorities, and migrants (Lai 2022).[1]

[1] Malaysia's Gini coefficient (2015: 43.03) and Palma ratio (2018: 1.969) exceed income equality global targets, indicating substantial income inequality (SDG for Malaysian States, n.d.).

R. H. Tan (✉)
Institute of Malaysian and International Study, National University of Malaysia, Bangi, Malaysia
e-mail: ranhan@ukm.edu.my

Y. B. Lee · M. Zainuddin
School of Media and Communication, Faculty of Social Sciences and Leisure Management, Taylors University, Selangor, Malaysia

N. H. A. Rahman
Faculty of Law, National University of Malaysia, Bangi, Malaysia

N. V. Roman et al. (eds.), *Social Inequalities Within and Among Countries*, Sustainable Development Goals Series, https://doi.org/10.1007/978-3-032-08741-6_17

Government initiatives to address inequalities have focussed on resource mobilisation and capacity building (Ministry of Economy 2021), particularly through the country's National Development Plans and the platform provided by the All-Party Parliamentary Group Malaysia on the Sustainable Development Goals (APPGM-SDG). Since 2019, the APPGM-SDG has facilitated dialogue between parliamentarians and the public, addressing local needs related to inequality reduction. To date, APPGM-SDG has implemented over 300 SDG-aligned solution projects at the grassroots level, specifically targeting marginalised communities. Furthermore, APPGM-SDG appoints researchers to evaluate the impacts of these initiatives.

The Central One team was appointed to evaluate solution projects implemented in parliamentary constituencies across Kuala Lumpur, Selangor, and Pahang (see Fig. 17.1). The team conducted an impact evaluation of 18 solution projects carried out between 2021 and 2022, culminating in the production of the Central Report, which details the projects' outcomes, challenges, and successes.

This chapter discusses the Central Report with specific reference to SDG Target 10.2 (Promote Universal Social, Economic and Political Inclusion). It examines how social interventions have contributed to reducing income inequalities, assesses their impacts, and explores the challenges encountered by beneficiaries.

17.2 Social Inequality, Income Inequality, and the B40 Community in Malaysia

Social inequalities arise from the negation and violation of equality (Therborn 2006) and encompass income inequality (Shin 2020). Carmo (2021) identifies three forms of social inequalities: vital inequalities, relating to life, health, and death (evident in high mortality rates amongst marginalised communities); resource inequalities, stemming from unequal access to social, economic, and cultural resources (such as education, income, and wealth); and existential inequalities, arising from systemic discrimination and limited opportunities for personal development.

In Malaysia, social inequalities persist as urban–rural and socio-economic divides continue to affect access to income, education, health care, and employment opportunities (Rongen et al. 2024; Khazanah Research Institute 2020). Vulnerable groups such as the Orang Asli and the indigenous communities of Sabah and Sarawak experience marginalisation, land rights violations, and limited access to essential services. Additionally, Malaysian women continue to face wage gaps, discrimination, and under-representation in leadership roles (Lee and Choong 2019).

The socio-economic status of Malaysian citizens is classified according to household income groups: T20 (top 20%), M40 (middle 40%), and B40 (bottom 40%). This classification enables the

Fig. 17.1 Central region FGD distribution

government to monitor economic progress, target assistance, and address inequalities (Samsudin and Nadzrulizam 2020). According to the Household Income and Basic Amenities Survey Report 2019, inequality, measured by the Gini coefficient, rose from 0.399 in 2016 to 0.407 in 2019, with the overall poverty rate reaching 5.6% in 2020 (Department of Statistics Malaysia 2020). The B40 population faced a significantly higher poverty rate of 18.4%, or one in five households (Singaravelloo and Salih 2022), and comprises 2.7 million homes, 44% in rural and 56% in urban areas, earning RM2,537.00 (USD 534) or less monthly.

This poverty rate highlights the need for focussed policy interventions aligned with international standards and treaties (Lim 2023). To enhance the social prosperity of the B40 community, it is crucial to provide access to education and skills. The B40 community can achieve social mobility through education, job training, health, migration, and individual efforts (Samsudin and Nadzrulizam 2020). To facilitate this, the Malaysian government has integrated the objectives of poverty alleviation and income inequality through the Malaysian Plans (Economic Planning Unit [EPU] 2016) (see Table 17.1).

Recognising the pivotal role of education in breaking the cycle of poverty, measures to improve access to quality education and skill development programmes were introduced (Nurul et al. 2023). Hence, the APPGM-SDG impact evaluation was conducted to assess the impact of the SDGs in the local contexts of the Central Region.

17.3 Methodology

17.3.1 Evaluation Framework and Concept

Globally, impact evaluations assess the effectiveness of interventions in generating sustainable positive changes, particularly in relation to poverty reduction and improvements in social welfare (Gertler et al. 2011). These assessments analyse changes in well-being linked to the intervention in order to determine its success (Bamberger 2000). Evaluations such as Mexico's influential conditional cash transfer (CCT) programme assessment have provided crucial evidence for informing policy decisions both locally and globally (Gertler et al. 2011).

The impact of solution projects on beneficiaries was evaluated using the four domains of deep, clear, high, and wide, as outlined in the MetaIntegral (2020) framework. To adapt the framework to APPGM-SDG's focus on localising the SDGs, two additional domains—SDG and Gender—were incorporated.

The Theory of Change (ToC) forms the foundation of the evaluation framework. As a purpose-driven framework, it outlines how and why desired changes are expected to occur within specific contexts, detailing how activities lead to outcomes and ultimately achieve long-term impacts (Daud and Ramli 2024).

Additionally, the evaluation framework is built on the concept of valorisation, which relates to the availability (of documents), accessibility (of activities to stakeholders), and usability (of

Table 17.1 The Malaysian plans

Malaysian plan	Duration	Focus
Tenth	2011–2015	• Education • Training • Entrepreneurship
Eleventh	2016–2020	• Economic well-being • Empowering human capital • Creating skilled, knowledgeable, and innovative human capital capable of meeting the demands of the industry
Twelfth	2021–2025	• Poverty • Covid-19 job training

materials by evaluators) of all solution projects (APPGM-SDG 2020).

17.3.2 Data Collection Methods and Sampling

The impact of a project on beneficiaries is assessed through document review, focus group discussions (FGDs), and site visits. The data collection instrument used is the Impact Identification Document, which is designed to capture the six domains of the evaluation frameworks. To gather data, the evaluators first examined all relevant documents related to the solution projects, including project proposals, monthly reports, and final reports. The document review compares the project's objectives, outcomes, and outputs, such as photographs, videos, testimonials, and reports. Following the document review and report writing, six projects were selected for the FGDs.

During the FGDs, the evaluators listened to the beneficiaries to verify findings from the document review and to incorporate additional observations for the Central Report. Five FGDs were conducted on-site in Lembah Pantai (Kuala Lumpur), Temerloh, and Bentong (Pahang), with one held via Zoom (Kuala Selangor, Selangor). Anonymity was maintained through identifiers (participants A, B, and C), and recorded discussions and documents were uploaded to Google Drive for transcription. At least two evaluators led each FGD using predetermined guidelines to ensure ethical practices. Within each hour-long session, they asked questions based on the six domains (see Table 17.2).

17.3.3 Data Analysis

The evaluators reviewed the dataset—including FGD notes, transcriptions, proposals, and reports—and employed summative content analysis (Hsieh and Shannon 2005) to identify keywords. They also recorded the frequency, extensiveness, and intensity of keywords mentioned by the participants. In addition, the specificity of the details provided, the consistency with which participants maintained their views, and their perceptions of importance were noted (see Krueger and Casey 2015).

17.4 Findings and Discussion

Three main factors are discussed: (1) social interventions to reduce income inequalities, (2) their impacts, and (3) challenges faced by beneficiaries.

17.4.1 Social Interventions to Reduce Income Inequalities and Their Impacts

The APPGM-SDG projects trained women from B40 communities to reduce income inequalities and alleviate poverty, in alignment with addressing income disparity and promoting economic growth within the framework of SDG 10. The impacts of these projects are evident through transformative outcomes evaluated across six domains.

Table 17.2 FGD interview protocol

Part	Probe/Key questions	Follow-up questions	Exit questions
Time	15 min	15 min	5 min
Questions	1. How did you get to know about the programme? 2. Why did you attend the programme? 3. What skills did you acquire? 4. If yes, how are the newly acquired skills benefiting you?	1. What knowledge about SDG did you gain? 2. Are you using SDGs in your work? 3. Would you like to provide additional information about the above? 4. Did you develop certain procedures/ structures through/because of the programme? 5. How do you feel about the programme?	1. Do you maintain contact with those you met at the programme?

Deep Impact:

Beneficiaries reported feeling empowered to generate income for their families, as the projects boosted their confidence, independence, and financial management skills. This empowerment fostered a shift towards economic independence and self-reliance, reflecting gratitude and a strengthened sense of agency. Addressing immediate financial needs, these outcomes contributed to poverty alleviation within B40 communities and aligned with the objectives of SDG 10.

Clear Impact:

Beneficiaries gained economic empowerment through skills development in areas such as baking, design, online entrepreneurship, floral arrangement, and horticulture. By equipping participants with practical and entrepreneurial skills, the projects enabled beneficiaries to pursue self-employment and promote inclusive economic growth, in line with SDG 10. This focus on skills development provided opportunities for economic independence beyond traditional societal roles.

Wide Impact:

The use of WhatsApp, social media platforms, and online marketplaces such as Shopee to network and establish businesses highlights the beneficiaries' adaptability in leveraging digital tools to overcome barriers to income generation. Whilst challenges related to comprehensive digital infrastructure remain, beneficiaries utilised digital networks to work towards reducing income inequalities within their communities. This impact demonstrates beneficiaries' intentions to expand cooperation with new networks.

High Impact:

Community engagement, participation, and sustainability were integral to the APPGM-SDG projects, ensuring that interventions were community-driven and responsive to economic vulnerabilities and gender-related challenges. Beneficiaries demonstrated ownership, agency, and autonomy by producing new work guidelines and engaging in inclusive decision-making processes, contributing to the goals of SDG 10.

SDG Impact:

Strategies implemented by beneficiaries closely aligned with the aims of SDG 10, focussing on reducing income inequalities and promoting inclusive economic growth regardless of background. This impact suggests that beneficiaries actively applied SDG principles in their work, fostering greater economic stability and access for all individuals within their communities.

Gender Impact:

Gender considerations were significant across the projects, as service providers, facilitators, and beneficiaries demonstrated a balanced representation of men and women. Three projects specifically targeted female beneficiaries, aiming to advance gender equality and women's empowerment. Two projects focussed on B40 Indian women. An intersectional analysis, considering gender, ethnicity, and class, reveals the distinct challenges faced by Indian women from lower socio-economic backgrounds (Saigaran and Thambiah 2023).[2] This layered perspective underscores the necessity for targeted policies and interventions to empower such groups. The FGDs highlighted how participants benefitted from entrepreneurial courses in tailoring, baking, and farming, which aligned with their cultural and geographical contexts. These initiatives facilitated financial independence, shifted gender dynamics, and fostered a sense of pride in providing for their families.

[2]In 2012, Malaysian Indians had the highest intra-ethnic income inequality (Gini coefficient of 0.44) compared to Bumiputera (0.42) and Chinese (Malaysian Indian Blueprint 2017). By 2019, the Gini coefficient for the Bumiputera increased from 0.385 to 0.389, Chinese from 0.411 to 0.417, and Indians from 0.382 to 0.411 (Department of Statistics Malaysia 2020). Malaysian Indian households tend to be patriarchal, and such norms dictate the roles and responsibilities of Malaysian Indian women (Hirschman 2016).

17.4.2 Challenges Faced by Beneficiaries

First: Income Inequalities

Many beneficiaries live below the poverty line, earn minimal incomes, and reside in rural areas or public housing. They often rely on informal, irregular employment to survive, face barriers to accessing proper education, marry early, or are sole breadwinners struggling to secure formal employment, educational opportunities, and health care. These challenges reflect global trends addressed by SDG 10, which seeks to reduce income inequality, promote equal opportunities, and dismantle systemic barriers to economic inclusion.

Second: Socio-Economic Status

Many single mothers act as sole breadwinners or homemakers and face limited access to childcare services. Some were unable to secure stable employment due to the economic downturn caused by the Covid-19 pandemic and the restrictions of the Movement Control Orders (MCOs) (Lee 2021). Similar socio-economic disparities have been exacerbated worldwide by the pandemic, highlighting the urgent need for enhanced social protection measures and inclusive economic policies.

Third: Access to Opportunities

Many beneficiaries expressed a strong desire to attend further training programmes to upskill and improve their financial circumstances. However, as economic and political power is largely centralised in Malaysia's major cities, many participants lacked the necessary funding to extend their training (Ahmad and Lee 2015). Those living in rural areas, who often lacked proper equipment or reliable Internet connections, found it particularly challenging to participate in online workshops. Globally, rural communities similarly face unequal access to resources and opportunities, an issue that SDG 10 explicitly seeks to address by promoting equal access to economic and social opportunities.

17.5 Conclusion

Inequality is intersectional; it overlaps and intersects across various identities and vulnerabilities. These project outcomes align with the SDG 10.2 objectives, which aim to mitigate income inequalities and enable everyone to thrive, irrespective of gender, race, religion, or economic standing. With a commitment to 'Leaving No One Behind', these projects have empowered women from the B40 community. Despite challenges such as low attendance, limited resources, and Covid-19, these initiatives had positive impacts. They enhanced beneficiaries' skills and confidence and fostered a sense of community and network building, contributing to the broader goals of socio-economic development and gender equality. As beneficiaries are keen to expand their knowledge and skills further, phase two of the projects should be introduced to help reduce income inequalities and alleviate poverty. This phase should include digital marketing, social media skills, and training in agriculture and entrepreneurship. Thus, a robust ToC empowers beneficiaries, customises interventions, ensures accountability, and creates a sustainable positive impact.

Acknowledgements The authors thank APPGM-SDG, beneficiaries, solution providers, Sunway University, and Universiti Kebangsaan Malaysia (UKM) for their support.

References

Ahmad M, Lee YB (2015) Media, alam sekitar dan golongan pinggiran. In: Azmawati AA et al (eds) Antologi esei komunikasi, vol 1. Teori, isu dan amalan. USM Press, Penang, pp 122–141

All-Party Parliamentary Group Malaysia for Sustainable Development Goals (APPGM-SDG) (2020) Annual report. Secretariat of APPGM-SDG, Kuala Lumpur

Bamberger M (2000) Opportunities and challenges for integrating qualitative and quantitative research. In: Bamberger M (ed) Integrating quantitative and qualitative research in development projects. The World Bank, Washington, DC

Carmo RM (2021) Social inequalities: theories, concepts and problematics. SN Soc Sci 1:116

Daud S, Rashila R (2024) Sustainable development goals (SDGs) and project impact evaluation in Malaysia. PSSM, Bangi

Department of Statistics Malaysia (2020) Household income and basic amenities survey report 2019. https://v1.dosm.gov.my/v1/index.php?r=column/cthemeByCat&cat=120&bul_id=TU00TmRhQ1N5TUxHVWN0T2VjbXJYZz09&menu_id=amVoWU54UTl0a21NWmdhMjFMMWcyZz09. Accessed 27 Feb 2024

Economic Planning Unit (EPU) (2016) Eleventh Malaysia plan. Strategy paper 2: elevating B40 households towards a middle-class society. Economic Planning Unit, Putrajaya

Gertler PJ, Martinez S, Premand P, Rawlings LB, Vermeersch CM (2011) Impact evaluation in practice. The World Bank, Washington, DC

Hirschman C (2016) Gender, the status of women, and family structure in Malaysia. Malays J Econ Stud 53(1):33–50

Hsieh H-F, Shannon SE (2005) Three approaches to qualitative content analysis. Qual Health Res 15(9):1277–1288. https://doi.org/10.1177/1049732305276687

Khazanah Research Institute (2020) Social inequalities and health in Malaysia: the state of households 2020 Part III. Khazanah Research Institute, Kuala Lumpur

Krueger RA, Casey MA (2015) Focus groups: a practical guide for applied research, 5th edn. Sage, London

Lai JY (2022) SDG for Malaysian states: reduced inequalities. https://sdg-for-malaysian-states-sdsn.hub.arcgis.com/pages/sdg-10-reduced-inequality. Accessed 27 Feb 2024

Lee YB (2021) Malaysian cinema in the new millennium: transcendence beyond multiculturalism. HKUP, Hong Kong

Lee H-A, Choong C (2019) Inequality in Malaysia: empirical questions, structural changes, gender aspects. J Southeast Asian Econ 36(3):329–354

Lim J (2023) T20, M40, B40 household update 2023. https://mefacademy.mef.org.my/Attachments/MSN20230731a.pdf. Accessed 18 Mar 2025

Malaysian Indian Blueprint (2017) Malaysian Indian blueprint. Malaysian Indian Blueprint Secretariat, Prime Minister's Office, Putrajaya

MetaIntegral (2020) Meta impact framework. https://www.metaintegral.com/. Accessed 5 Feb 2023

Ministry of Economy (2021) Sustainable development goals. https://www.ekonomi.gov.my/en/sustainable-development-goals. Accessed 12 Jan 2024

Nurul HAR, Redwan Y (2022) Children's rights to "zero hunger" and the execution challenges during the COVID-19 crisis. Hasanuddin Law Rev 8(2):139–159

Nurul HAR, Lee AYB, Tan RH (2023) Implementation of APPGM-SDG solution projects and impact evaluations in Malaysia's central region. J Tech Educ Train 15(3):182–191

PATHFINDERS (2023) A call to action to save SDG10. https://www.sdg16.plus/resources/a-call-to-action-to-save-sdg10/. Accessed 25 Apr 2024

Rongen G, Ahmad ZA, Lanjouw P et al (2024) Regional and ethnic inequalities in Malaysian poverty dynamics. J Econ Inequal 22:101–130

Saigaran NG, Thambiah S (2023) Malaysian Indian women's education and their marginalisation: an intersectional analysis. In: Nagappan R, Mukherjee H (eds) Malaysian Indians and education: reimagined development opportunities. Routledge, Oxon, pp 100–119

Samsudin HB, Nadzrulizam AA (2020) Relationship between B40 household income and demographic factors in Malaysia. Int J Innov Technol Explor Eng 10(2):113–117

Shin KY (2020) A new approach to social inequality: inequality of income and wealth in South Korea. J Chin Sociol 7:17

Singaravelloo K, Salih K (2022) Cities, urbanisation and infrastructure. In: Rasiah R, Salih K, Cheong KC (eds) Malaysia's leap into the future: the building blocks towards balanced development. Springer, Singapore, pp 237–263

Therborn G (2006) Meaning, mechanisms, patterns, and forces: an introduction. In: Therborn G (ed) Inequalities of the world: new theoretical frameworks, multiple empirical approaches. Verso, London, pp 1–58

United Nations Development Programme (UNDP) (2023) Sustainable development goals. https://www.undp.org/sustainable-development-goals. Accessed 18 Mar 2025

Wilber K (2000) A theory of everything: an integral vision for business, politics, science, and spirituality. Shambhala, Boston

Conclusion

18

Shewli Kumar, Rashila Ramli, Nicolette Roman, and Beatrix Schwarzer

> Inequality not just the less privileged but also the society as a whole by undermining social cohesion and economic stability
>
> *– Amartya Sen (1992)*

Abstract

This chapter consolidates the key concepts presented in the book. It reiterates the need to explore and understand inequalities within and amongst countries from a social justice and intersectional framework, which informs the core arguments in all the chapters. The need for social work education and interventions are highlighted and explored in the four countries to address these inequalities. The country-specific case studies enhance the need to explain and research social inequalities from a localised and contextual approach and encourage the need for interventions which enable social equality. This chapter finally emphasises the need for transnational research and social work interventions to address inequalities within and amongst countries.

Keywords

Inequalities · Social work interventions · Intersectional · Transnational research

18.1 Introduction

This book focuses on Sustainable Development Goal (SDG) 10, which aims to address inequalities within and amongst countries—inequalities that are deeply embedded in the social structures of societies worldwide. It places particular emphasis on four countries: Malaysia, Germany, South Africa, and India. These inequalities operate across local, national, and global levels, enabling the powerful to dominate global decision-making, exercise exclusionary control over the world's resources, and deepen disparities both between and within nations. We live in an interdependent world, where global shifts—such as wars, conflicts, climate change, and disasters—affect everyone, albeit to varying degrees. Across all countries, there is clear evidence of institutional failure, as governments struggle to navigate rapid changes in the global economy and address the

S. Kumar (✉)
School of Social Work, Tata Institute of Social Sciences, Deonar, Mumbai, India
e-mail: shewli4@tiss.ac.in

R. Ramli
Universiti Poly-Tech Malaysia, Kuala Lumpur, Malaysia
e-mail: rashilaramli@uptm.edu.my

N. Roman
University of the Western Cape, Bellville, South Africa
e-mail: nroman@uwc.ac.za

B. Schwarzer
University of Frankfurt of Applied Sciences, Frankfurt Main, Germany
e-mail: schwarzer@fb4.fra-uas.de

N. V. Roman et al. (eds.), *Social Inequalities Within and Among Countries*, Sustainable Development Goals Series, https://doi.org/10.1007/978-3-032-08741-6_18

widening inequalities that are emerging (Conceicao 2024).

There has been considerable debate about the efficacy of the SDGs and criticism of their weak implementation strategies, particularly at the local level (Androff and Damanik 2024). It must be recognised that the SDGs can only be achieved if local communities—especially those living in marginalised circumstances—are included in discussions and actively participate in developing relevant implementation strategies. At their core, the SDGs aim to create a better and more equitable world for everyone. The case studies from the four countries underscore the unique diversities and inequalities present in each context, highlighting the importance of addressing and incorporating these factors into the implementation of SDG 10. At the same time, localisation is intricately connected to global processes marked by negative developmental trends such as wars, trade barriers, increasing human rights violations, and the denial of a level playing field for decision-making, all of which further deepen the inequalities amongst nations.

Social work, social services, and community engagement play a vital role in advancing the SDGs, promoting a human rights perspective within development (Jayasooria 2016; International Federation of Social Workers [IFSW] 2021; Seth 2018; Nikku and Pulla 2014; Kaltenborn et al. 2020; Pandey and Kumar 2018). Moreover, development cooperation calls for 'services that are responsive to people's needs' (Federal Ministry for Economic Cooperation and Development [BMZ] 2017). Thus, the SDGs also highlight the need for adjustments in social work and social science curricula (Pandey and Kumar 2018; IFSW 2021). Whilst SDG monitoring systems rely heavily on indicators, statistics, and randomised trials (United Nations Statistical Commission 2017), social work and applied social science must grapple with complex social problems, cultural practices, and multilayered social interventions (Kaufmann 2012).

With this as a background, SDG 10—Reduce Inequalities Within and Among Countries—became the common ground for a collaborative research-based project amongst the four universities of South Africa, India, Malaysia, and Germany. As the World Social Science Report (United Nations Educational, Scientific and Cultural Organization [UNESCO] 2016) demonstrates, research on inequalities and social justice draws from numerous subfields, including political science, sociology, gender studies, social work, development studies, developmental psychology and family studies, criminology, and law. Social work and applied social science are inherently multidisciplinary, engaging young women and men as emerging researchers, global citizens, and agents of change.

Social work and applied social science bring a unique perspective to the study of inequalities, as they do not focus solely on 'problems' but also illuminate coping strategies at multiple levels (individuals, families, and communities) and aim to design adequate social interventions (political, economic, legal, welfare, care, etc.). At the same time, evidence-based practice is gaining increasing importance in social work, making the application of research to practice and policy settings essential.

For the implementation, monitoring, and evaluation of progress on SDG 10, it is crucial to draw on the expertise of social work and social scientists. The key focus areas include social, economic, and political inclusion of all (10.2); the elimination of discriminatory laws, policies, and practices (10.3); migration and the mobility of people (10.7); as well as support for development efforts (10.A). Methodologically, approaches should emphasise people-centred, community-oriented designs (Schenck et al. 2015; Isaacs et al. 2019; Gutiérrez and Gant 2018; Malata-Silva 2017; Kumar 2017) and closely link research to practice (Uggerhøj 2019; Austin and Isokuortti 2016; Schenck et al. 2015). By joining forces, the partner institutions aim to integrate perspectives from both the Global North and Global South in the study of social inequalities, discriminatory practices, social interventions, and personal, family, and community resilience..

18.2 (In)equalities, Intersectionality, and Social Justice

Understanding inequalities within and amongst countries is challenging, as the world is no longer seamlessly divided into a Global South and Global North. Local inequalities reflect a mix of traditional socio-cultural and economic power dynamics, with privileged groups set against marginalised and discriminated populations. Global shifts, driven by the neoliberal restructuring of economies and political systems, have exacerbated local inequalities, deepening poverty, deprivation, and destitution for many. As Androff and Damanik (2024) note, rising temperatures, heatwaves, and droughts are symptoms of rapid climate change, conditions that intensify the exploitation of vulnerable populations, even as wealth and profiteering amongst the elite continue to grow. Inequality reinforces existing power dynamics and social divisions, acting as a barrier to political solutions. Global partnerships are increasingly threatened by rising authoritarianism, disinformation, and xenophobia, which can destabilise democratic institutions, free speech, and human rights.

This book therefore adopts a conceptual framework that explores social inequality, intersectionality (Crenshaw 1989), and the pursuit of social justice intersect with SDG 10. By examining the multiple layers of social inequality present in all societies, it emphasises the impact of overlapping identities—including gender, race, ethnicity, disability, and socio-economic status. The discussion then turns to the concept of intersectionality, an essential lens for understanding the interconnected systems of advantage and discrimination that shape relationships between individuals and groups.

The complex relationship between social justice and inequality is further explored, demonstrating how structural inequalities perpetuate unjust patterns, hinder the realisation of fundamental human rights, and obstruct efforts to build truly just and inclusive societies. The text highlights that achieving meaningful social justice requires addressing the root causes of inequality, such as systemic bias, unequal power distribution, and inequitable access to resources and opportunities. Reducing inequalities demands a multifaceted approach that takes into account the complex intersections of various forms of discrimination and privilege across multiple levels—from families and communities to state institutions.

Whilst inequalities have long existed both within and between countries, they become unjust when they create barriers that prevent certain groups from accessing the resources needed to live a fulfilling life with dignity and without fear. Social justice offers a framework for promoting the equitable distribution of resources and opportunities, recognising the dynamic, ever-evolving nature of social power hierarchies and the capacity of individuals and communities to overcome adversity.

18.3 Relevance of Social Research, Social Work Education, and Interventions for Transnational Cooperation and Implementation of SDG 10

The discourse on the SDGs and the need for genuine implementation cannot be confined to a single-country approach. Transnational research, cooperation, and the exchange of ideas are essential for building a shared knowledge base that can inform possible ways forward and foster global cooperation. Chapters 1 and 2 (of Sect. 2) explored the efficacy of social work and social sector interventions, followed by Chap. 3 which highlighted the relevance of the SDGs for social research.

Chapter 1 provided an overview of the four countries and their profiles (Germany, India, Malaysia, and South Africa). It mapped social work education and situated the diversities of country context and curriculum, including the socio-economic context, laws and policies, and education and training. It also discussed the similarities and differences between the four countries and drew conclusions from a transnational perspective. Chapter 2 built on the first one, reaffirming the central promise of the 2030 Agenda for Sustainable Development: to 'leave no one behind'. It highlighted how social work as a global profession has emerged in response to local problems stemming from inequality, conflict, poverty, war, and other global phenomena. Social work is closely linked to values such as human rights, social justice, and equality. Drawing on qualitative transnational research, the chapter showcased the diversity of practices and methods employed by social work practitioners in the four countries, shaped significantly by the nature of each welfare state and its socio-economic conditions. Together, these two chapters highlighted the potential of a transnational approach to addressing social inequalities both within and amongst countries. They provided valuable insights for social work educators, practitioners, and students, offering a lens through which to understand the diversity of societies, welfare states, and the global realities of inequality and discrimination.

Research is conducted across a wide range of disciplines and serves as the cornerstone for accumulating and disseminating valuable knowledge. Chapter 3 introduced sustainability research focussed on addressing reduced inequalities, which plays a critical role in creating a more just and equitable world. In a time when ecosystems and living conditions are being damaged on a massive scale, sustainability research is becoming increasingly important both nationally and internationally. This expansion has accelerated particularly since the introduction of the SDGs. Research advances knowledge by meticulously converting collected information into new insights, and the dissemination of this knowledge supports societal progress and promotes a deeper understanding of how we perceive and interact with the world, thereby enhancing overall development. This chapter underscored the importance of research in addressing SDG 10 by focussing on social intervention. To explore the relevance of research, the authors examined several targets under SDG 10, including investigating trends in inequality and identifying its components; assisting in the formulation of policies relevant to at-risk communities; raising awareness; and tracking progress through the effective use of resource allocation. The authors aimed to link these targets by scrutinising the analytical tools used to address social inequalities across countries and by applying intersectionality as a key framework.

Whilst the book addresses perspectives, approaches, and strategies for implementing the SDGs—and SDG 10 in particular—the next section presented case studies from the four countries, addressing some of the key indicators of SDG 10. It emphasised the need for locally conceptualised, participatory social interventions to tackle specific inequalities, including income inequality and livelihoods, discriminatory laws and policies, and the orderly, safe, regular, and responsible migration and mobility of people. Throughout all the chapters, gender is used as an underlying and explicit domain to highlight the intersectional discrimination faced by women.

18.4 Need for Addressing Discriminatory Laws and Policies

According to the Convention on the Elimination of Discrimination Against Women (Convention on the Elimination of Discrimination Against Women [CEDAW] 1979), all countries need to incorporate the principle of equality between women and men into their legal systems, abolish all discriminatory laws, and adopt appropriate legislation prohibiting discrimination against women. Similarly, SDG 5 clearly outlines the requirement for countries to end all forms of discrimination against women and girls everywhere, encourage participation at all levels, and eliminate all forms of violence against women. The two case studies presented in chapters seven and eight from India reveal that this is far from a reality, as patriarchy is embedded in all systems and institutions, creating multiple barriers for women, particularly those from marginalised communities, in accessing their rights.

The first case study on a tribal widow's bid for land in the eastern state of West Bengal demonstrated that the intersection of two laws may render rural, agrarian women as 'double beneficiaries'. Single women, such as unmarried daughters, or those who are deserted, divorced, or widowed, who are most in need of land, may become landowners by virtue of their labour and the inheritance principle, if both laws complement each other.

The second case study from India explored the state of Indian health insurance policies concerning their coverage of female-specific healthcare needs, through a comprehensive document analysis of publicly accessible policy documents from both public and private health insurers. The findings revealed major gaps in risk and cost coverage across three essential female-specific healthcare concerns: maternity care, gynaecological health, and mental well-being, highlighting discriminatory policies in the Indian healthcare system. This underscores the urgent need for reform in health insurance policies, ensuring equitable and justiciable access to health care for women. Social interventions, including fostering collaboration between law-abiding government and private sectors for developing effective coverage plans tailored to the specific health needs and social conditions of women, as well as greater investment in the healthcare system to enhance service delivery and reduce out-of-pocket expenses (OOPE), are essential.

18.5 Challenges and Opportunities for Social Interventions for International Migrants

Migration is a global phenomenon shapedby various socio-economic and political factors that affect people's lives. Individuals may migrate to study, due to economic conditions, to reunite with family, or—more recently—to escape persecution arising from conflict, war, and disasters. According to the United Nations Population Division Report No.13, UNDESA (2025), the number of international migrants in 2024 stood at 304 million, a figure that has nearly doubled since 1990. Europe hosted the largest number of international migrants, with Germany ranked second as a leading destination, hosting 16.8 million in 2024 (United Nations Department of Economic and Social Affairs [UNDESA] 2025).

The case studies in chapters nine and ten on migration from Germany and Malaysia reveal both similar and diverse situations. However, the common threads are the challenges that refugees face as they make precarious journeys to safer places, attempt to adjust to new cultures and social norms, and work to overcome trauma and distress—often whilst facing hostility from local populations who may view them as problematic, a strain on scarce resources, or unfair dependents. Unless the governments of destination countries establish clear mandates as set out in international law and allocate sufficient resources for migrant populations, the work of social workers and social interventionists becomes extremely challenging. These case studies illustrate the urgent need for social interventions to support international migrants in assimilating and building better lives in their destination countries.

The German case study explored the migration of Afghan refugees, detailing their circumstances in the new country and the efforts to create enabling conditions through policy. According to the Office of the United Nations High Commissioner for Refugees (UNHCR), Afghans form the third-largest refugee group globally, after Syrians and Ukrainians. Of the 5.7 million Afghan refugees worldwide, nearly 400,000 are currently in Germany, with approximately one-third being women. Facilitating safe and responsible migration through policy is one of the SDG 10 targets to reduce inequalities (10.7). This study examines the life situation of Afghan evacuees after 2021 and discusses how German policies and regulations respond to the needs of the Afghan diaspora.

The Malaysian case study illustrated the need to engage and work with Rohingya refugees from Myanmar. Although Malaysia is a signatory to international commitments, the absence of specific legal measures within domestic law exposes Rohingya refugees to exploitation and denies them access to fundamental protection mechanisms. This includes Rohingya children, who are vulnerable to various forms of exploitation, including trafficking, child labour, child marriage, and increased risk of arrest or detention. Social workers provide individualised support, advocacy, and collaboration with stakeholders to safeguard the rights and well-being of these vulnerable children. By addressing systemic gaps and promoting community empowerment, social workers can help create more protective and inclusive living environments for Rohingya refugee children, enabling them to thrive and reach their full potential.

18.6 Increasing Social Inequality and Precarity in Livelihoods

The World Inequality Report (WID.WORLD 2024) explains how income inequality has been reducing, but there are stark regional and country-specific disparities. Based on the World Inequality Report (Stanley 2022), the International Monetary Fund (IMF) analyses that the poorest half of the world's population owns only €2,900 (in purchasing power parity) per adult, whilst the top 10% own 190 times more. This set of large data hides the extreme forms of localised and country-specific income and livelihood inequality. Groups and communities living on the margins of society have the least access to income and other assets to escape poverty, and big data is unable to highlight these specificities. This is amply demonstrated by some of the country-level case studies (chapters eleven, twelve, and thirteen) in the book, such as the chapter on the Seharia, a particularly vulnerable tribal group in Rajasthan, India. The Seharia struggle due to a lack of resources, employment opportunities, and nutrition. They have historically depended on forests; however, this is changing due to shifts in forest-related policies and composition. The study findings revolve around forest degradation, land grants and grabs, and bonded labour, which significantly impact the community's livelihood in the state. The findings reflect that the Seharia face socio-economic exclusion, making the SDG of reduced inequality and inclusion remain elusive for them. Another study based on the livelihoods of the Gond tribe in India reiterates the need for gender- and tribe-specific approaches to strategies addressing social inequality. Similarly, the chapter on the livelihood challenges of female-headed households in Evaton, South Africa, shows that inequalities are felt at the household level along race, gender lines, and particularly geographical location. The findings indicate that poverty cuts deep in these households, and the only stable source of income is the social security grant, child support grants, and stipends obtained by women from their voluntary work in non-governmental organisations (NGOs) in their local communities. All the women were unemployed and had limited income opportunities. They described their living conditions as a 'difficult life'.

These case studies also sharpen our understanding and approach to inequality, which is social, structural, and intersectional. Just measuring income inequality and its indicators is limited and fails to capture all aspects of social inequality.

18.7 Equal Opportunity and Its Promotion to Address Social Inequality

Whilst the country-level case studies highlight specific deprivations, there is also a need to address these through social interventions that are sensitive to local realities. To achieve the goals of SDG 10, the local and global must be connected and contextualised. Social workers and social interventionists from various disciplines have long engaged with inequalities and emerging challenges such as poverty, youth gangs, rural exclusion, and women's and girls' lack of access to education, land, and other economic resources. Some of the case studies underscore the need for such social interventions to create enabling conditions for populations that are 'left behind' in national development efforts.

Specific case studies (chapters fourteen, fifteen, and sixteen) from South Africa and Malaysia share experiences of addressing the needs of certain population groups. One ethnographic study from South Africa sought to gain deeper insight into the experiences of Black African and Coloured boys (aged 10–16) awaiting trial in a Child and Youth Care Centre (CYCC). The main research methods employed included individual interviews, focus group discussions, and participant observation with the boys as key participants. Based on the collected data, the boys' experiences and narratives revealed how intergenerational poverty shaped their lives, pushing them to seek social, economic, and physical protection by joining street gangs—pathways that often led them into criminal and violent experiences.

The case study with adolescent girls in farming communities in the Western Cape, South Africa, brings forth their challenges for accessing education and possible social interventions. Facing intersecting inequalities rooted in gender, poverty, geography, and cultural norms, these girls have limited educational access and empowerment opportunities. An action research study explains how contextually tailored non-formal education programmes can serve as transformative interventions to address the social inequalities faced by this marginalised population. The findings revealed that the programme enhanced participants' confidence, communication abilities, and self-expression, strengthened family relationships, and disrupted patriarchal gender norms that had marginalised their voices. The study highlighted the significant role of mentorship in providing growth and development for girl-children beyond what families alone can offer.

Similarly, another case study explored the lives of children in several rural and urban poor areas in peninsular Malaysia who have historically been excluded from kindergarten programmes. Such gaps prompted social intervention through community-based kindergarten efforts designed to bridge inequalities and develop innovative programmes to address the intersectionality of these inequalities through social action.

These experiences showcase the importance of local and country-specific interventions as a core strategy for achieving the SDGs, particularly SDG 10. Often, populations and groups living on the margins are lost in broad data counts and high-level analyses of social inequality. It is imperative that a bottom-up approach, focussed on identifying the most deprived and marginalised, ensures a deeper and more meaningful understanding of inequality and effectively informs global efforts to reduce it.

18.8 Monitoring and Evaluation (M&E) of SDG 10: A Need of the Hour

The Global Agenda 2030, commonly known as the SDGs, was adopted by UN member states in 2015 as a universal call to action to address pressing global challenges by 2030. Effective M&E is crucial for tracking progress, ensuring accountability, and informing policy decisions. The last two chapters (Chaps. 16 and 17) examine the M&E system employed by the United Nations (UN) for the SDGs, with a focus on its implementation at the global, regional, and national levels.

At the global level, the United Nations Statistical Commission and the Inter-agency and Expert Group on SDG Indicators (IAEG-SDGs) play pivotal roles in data collection and reporting. Regionally, organisations such as UN ESCAP (United Nations Economic and Social Commission for Asia and Pacific) facilitate data harmonisation and regional assessments. Nationally, countries integrate SDG indicators into their development plans, utilising national statistical offices and collaborating with civil society. Voluntary National Reviews (VNRs) are reporting mechanisms through which member states can showcase their accomplishments.

The discussion underscores the challenges and opportunities in achieving comprehensive, accurate, and timely evaluations. The Mid-Term Review in 2023 indicated that inequality amongst countries has increased. By analysing the multi-tiered approach of the UN M&E system and the Impact Ranking System, Chapter 16 in this section outlines best practices and future directions in understanding inequality. Chapter 17 further illustrates the work done in Malaysia to embed the implementation, monitoring, and evaluation of the progress of the SDGs. It provides clear direction on how social interventions reduce income inequalities, their impacts, and the challenges beneficiaries face.

18.9 Conclusion

This book is the result of the collaborative efforts of four universities working together to research, converse, dialogue, and debate on SDG 10—Inequalities Within and Among Countries. These universities have committed to working with communities and groups belonging to marginalised sections of their countries, which has illuminated the deep and entrenched inequalities they face. Over the four-year period of collaboration amongst academics, young research scholars, and social workers, it became evident that no pathway towards a more equal world can be achieved without localising the SDGs, particularly SDG 10. The active involvement of local stakeholders is essential for the global fulfilment of these broad and deep goals. Simultaneously, transnational engagement, conversations, and research collaborations broaden the scope for achieving these objectives. This book is a modest attempt in this direction.

There are four major findings that can be drawn from Sects. 2, 3, and 4. First, the cases illustrate the interconnectivity between the SDGs. In the case of the Seharia, a particularly vulnerable tribal group in Rajasthan, India, and the case of Coloured boys in South Africa, the prominence of the SDG on gender (in)equality was apparent. In both cases, the issue of inter-categorical intersectionality concerning gender, race, and locality demonstrated the impact that leads to greater inequality. Second, in all cases, institutions and the presence or absence of partnerships can significantly affect social inequality. The partnership of institutions, such as civil societies and UNHCR in Afghanistan and Malaysia, has impacted the education of migrant children as well as displaced adults. Third, social justice can be achieved through social intervention, as shown in the chapters on rural-based kindergartens for children in Malaysia, as well as through public–private institutional cooperation in securing much-needed health insurance for the poor in India. Finally, this study found that transnational collaboration amongst scholars brought different perspectives and interpretations to light. Inter-generational exchanges between scholars also enhanced the findings of this study.

The SDGs are a roadmap to a better world, and their aspirational slogan 'leave no one behind' represents an uphill task that requires the full commitment of people, communities, and governments across the globe.

References

Androff D, Damanik J (eds) (2024) The routledge international handbook of social development, social work and the sustainable development goals. Routledge, Oxon

Austin MJ, Isokuortti N (2016) A framework for teaching practice-based research with a focus on service users. J Teaching Soc Work 36(1):11–32. https://doi.org/10.1080/08841233.2016.1129931

Conceicao P et al (2024) Breaking the gridlock: reimagining cooperation in a polarized world. Human Development Report 2023/24. United Nations Development Programme, New York

Crenshaw K (1989) Demarginalizing the intersection of race and sex: a black feminist critique of antidiscrimination doctrine, feminist theory and antiracist policies. Univ Chic Leg Forum 1989(1):139–167

Gutiérrez LM, Gant LM (2018) Community practice in social work: reflections on its first century and directions for the future. Soc Serv Rev 92(4):617–646. https://doi.org/10.1086/701640

IFSW (2021) Social Work and Sustainable Development. https://www.ifsw.org/social-work-and-the-united-nations-sustainable-development-goals-sdgs/

IFSW (2021) 2020 to 2030 Global agenda for social work and social development framework: co-building inclusive social transformation. https://www.ifsw.org/2020-to-2030-global-agenda-for-social-work-and-socialdevelopment-framework-co-building-inclusive-social-transformation/

Isaacs SA, Roman NV, Savahl S (2019) An exploration of the family resilience needs of a rural community in South Africa: a sequential explanatory mixed methodological study design. Curr Psychol 38(6):1634–1648. https://doi.org/10.1007/s12144-017-9722-5

Jayasooria D (2016) Sustainable development goals and social work: opportunities and challenges for social work practice in Malaysia. J Hum Rights Soc Work 1(1):19–29. https://doi.org/10.1007/s41134-016-0007-y

Kaltenborn M, Krajewski M, Kuhn H (eds) (2020) Sustainable development goals and human rights. Springer, Cham

Kaufmann F-X (2012) Konzept und formen sozialer intervention. In: Albrecht G, Groenemeyer A (eds) Hand-

buch soziale Probleme. VS Verlag für Sozialwissenschaften, Wiesbaden, pp 1285–1305
Kumar S (2017) The gendered terrain of maintenance for women: enmeshed inequalities in culture and law. Scholarink, New Delhi
Malata-Silva J (2017) Community organizing participatory action research (COPAR) in curriculum development. ASEAN J Commun Engagem 1(1):29–40
Nikku BR, Pulla V (2014) Global agenda for social work and social development: Voices of the social work educators from Asia. Int Soc Work 57(4):373–385. https://doi.org/10.1177/0020872814527633
Pandey UC, Kumar C (2018) An SDG compliant curriculum framework for social work education: issues and challenges. In: Leal Filho W (ed) Implementing sustainability in the curriculum of universities: approaches, methods and projects. Springer, Cham, pp 193–206
Schenck R, Hanna N, Huma L (2015) Community Practice. Unisa Press.
Sen A (1992) Inequality re-examined. Clarendon Press, Oxford
Seth N (2018) Social services and the SDG. Available online at https://unitar.org/sites/default/files/media/file/social_services_and_the_sdgs-April2018.pdf.
Stanley A (2022) Global inequalities. Finance Development Magazine, IMF. https://www.imf.org/en/Publications/fandd/issues/2022/03/Global-inequalities-Stanley. Accessed 18 Apr 2025
Uggerhøj L (2019) What is practice research in social work—definitions, barriers and possibilities. Soc Work Society 17(2)
UNESCO (2016) World social science report 2016. Challenging inequalities: pathways to a just world. Edited by United Nations Educational, Scientific and Cultural Organization (UNESCO). International
United Nations Department of Economic and Social Affairs (2025) International migrant stock: key facts and figures, population division, UN/DESA/POP/2024/DC/No.13, Jan 2025. https://www.un.org/development/desa/pd/content/international-migrant-stock. Accessed 7 April 2025
United Nations Statistical Commission (2017). https://unstats.un.org/UNSDWebsite/statcom/48. Accessed 2 March 2020
WID.WORLD (2024) World inequality report 2024: 10 facts on global inequality in 2024. https://wid.world/news-article/10-facts-on-global-inequality-in-2024/. Accessed 18 Apr 2025
UN Women (1979) Convention on the elimination of all forms of discrimination against women (CEDAW) https://www.un.org/womenwatch/daw/cedaw/. Accessed 7 April 2025

Index

N. V. Roman et al. (eds.), *Social Inequalities Within and Among Countries*, Sustainable Development Goals Series, https://doi.org/10.1007/978-3-032-08741-6

The manufacturer's authorised representative in the EU is Springer Nature Customer Service Centre GmbH, Europaplatz 3, 69115 Heidelberg, Germany. If you have any concerns regarding our products, please contact ProductSafety@springernature.com

Printed and bound by CPI Group (UK) Ltd, Croydon, CR0 4YY
17/09/2026
02218463-0002